Value Management

The value management benchmark:

A good practice framework for clients

and practitioners

Steven Male, University of Leeds

John Kelly, Heriot-Watt University

Scott Fernie, Heriot-Watt University

Marcus Grönqvist, University of Leeds

Graeme Bowles, Heriot-Watt University

Thomas Telford

EPSRC

HERIOT-WATT
UNIVERSITY
Edinburgh

Published by Thomas Telford Publishing, Thomas Telford Ltd, 1 Heron Quay, London E14 4JD.
URL: http://www.t-telford.co.uk

Distributors for Thomas Telford books are
USA: ASCE Press, 1801 Alexander Bell Drive, Reston, VA 20191-4400, USA.
Japan: Maruzen Co. Ltd, Book Department, 3-10 Nihonbashi 2-chome, Chuo-ku, Tokyo 103
Australia: DA Books and Journals, 648 Whitehorse Road, Mitcham 3132, Victoria

First published 1998

A catalogue record for this book is available from the British Library

ISBN: 0 7277 2729 X

Typeset by Kneath Associates, Swansea
Printed and bound in Great Britain by Halstan & Co Ltd, Amersham, Bucks

Contents

Foreword

Any construction project should only be commissioned following a careful analysis of need since failure to think through project requirements will almost certainly cause problems for subsequent design and construction stages. For that reason, the Construction Industry Board recommends that value management be incorporated as an integral part of the construction process.

Value management is becoming increasingly important as projects are being correctly assessed on the basis of life cycle returns, such as schemes arising from the Private Finance Initiative. In these cases in particular, the use of value management highlights the overriding importance of an early and comprehensive project briefing involving all stakeholders. The prize in terms of improved value-for-money can be significant, but of almost greater importance is the benefit in terms of team building. Hidden agendas are exposed and dealt with, allowing everyone to get on with the prime job of delivering a quality project with minimal changes.

The approach taken to value management will undoubtedly vary depending on the procurement route chosen, but generally it is a technique that can be usefully applied on any construction project. Contrary to popular belief it is not just a matter of brainstorming and problem solving. To work properly it requires a structured methodology, a good facilitator, and consensus from all the stakeholders about the needs and objectives of the project. It is also vital that participants bring the commitment and authority to implement the actions that arise.

I welcome this good practice framework in particular for the guidance it provides by highlighting specific value points in the project life cycle and detailing practical tools and techniques that can be used by value managers and facilitators.

Richard Baldwin,
Managing Director of Special Projects at Alfred McAlpine plc
Productivity and Cost Improvement Panel, Construction Industry Board

Acknowledgements

The authors would like to thank the following people and organisations for their help and assistance in producing this value management framework.

Eric Adam, Retired
Michael Adams, Adams & Associates
David Adlington, The Lord Chancellor's Department
Boris Arratia, US Army Industrial Operational Command
Clive Bone, Bone & Robertson
Catherine Boothroyd, Associate Dierctor of Risk Management, Gardiner & Theobald
John Bryant, President, Value Management Associates
Tony Burton, Partner, Gardiner & Theobald
John Bushell, John Bushell Value Management
Harvey Childs, Architect, Office of Financial Management
Lex Clark, Clark Management and Engineering Pty Ltd
John Connaughton, Davis Langdon Consultancy
Mike Dallas, Davis Langdon & Everest
Guido Dandri, Professor, University of Genoa
Peter Dark, VSI
Winston Davies, Jaguar Cars Ltd
Brian Dawson, Value Systems Pty Ltd
Richard Dinham, SJPH Design Partnership
Tom Dominy, Tom Dominy Soloution Facilitators
Derek Drysdale, Railtrack
Howard Ellegant, Howard Ellegant Associates
Clive Evans, Principal, LEC Quantity Surveying Inc.
Brian Farmer, Capital Insight Pty Ltd
Peter Filby, Deakin University
Drummond Graham, Thomson Bethune
Colin Harris, Ove Arup and Partners
Elizabeth Heider, Senior Project Manager, Hanscomb
Dee Higer, Staff, Corporate Project Administration, Mitsubishi, Motor Manufacturing of America, Inc.
Robert Hunter, Director, Technical Resources Pty Ltd
Masaru Imau, Crosfield Electronics
Sunny Kawamura, Crosfield Electronics
Geoff Kier, Balfour Beatty
Thomas King, Senior Project Manager, Lewis & Zimmerman Associates, Inc.
Chris Laird, NSW Department of Public Works and Services
Mathew Locke, Bovis
Professor Luigi Maffei, University of Pisa
Frank Mase, Value Process Engineer, Pratt & Whitney
Francis McLean, US Bureau Of Reclamation
Gary Milligan, NSW Department of Public Works and Services
Giovanni Minelli, Ansaldo
Michael Morrison, Value Management Consulting, Inc.
Brian Peachey, Franklin and Andrews
Martyn Phillips, Value Management International
Peter Popper, Higgs & Hill
Russell Poynter-Brown, Dearle and Henderson

Ross Prestipino, Director, The Australian Centre For Value Management Pty Ltd
Associate Professor Roy Barton, University of Canberra
Martyn Quarterman, Appleyards
James Rains, Supervisor Value Management, General Motors
Elizabeth Randall, Schal
James Reid, Osprey Project Management
George Riek, Vice President, Camp Dresser & McKee Inc.
Paul Riley, Jaguar Cars Ltd
Diana Sands, Development Manager, Group Technical Services, BAA plc
Larry Shillito, Eastman Kodak Company
Nigel Standing, Southern Water
Robert Stewart, President, Consulting Value Specialists Inc.
Luigi Tenenti, Valor Team
Michel Thiry, Bovis
Declan Tierney, Tierney Page Kirkland Pty Ltd
Harry Townley, Balfour Beatty
Brian Vawser, CDG International
Dr William H. Copperman, President, Copperman Associates in Value
    Engineering Inc.
Roy Woodhead, Oxford Brookes University
Mike Yardley, London Underground Ltd

Abbreviations used in this report

AIA	American Institute of Architects
BRE	Building Research Establishment
BSRIA	Building Services Research and Information Association
CIB	Construction Industry Board
CIRIA	Construction Industry Research and Information Association
CSF	critical success factor
CUP	HM Treasury Central Unit on Procurement
FSD	final sketch design
ICE	Institution of Civil Engineers
LCC	life cycle costing
OSD	outline sketch design
PEP	project execution plan
POE	post-occupancy evaluation
PPE	post-project evaluation
RIBA	Royal Institute of British Architects
SWOT	strengths, weaknesses, opportunities and threats
T/C/Q	time/cost/quality
VM	value management
VMCP	value management change proposals
VMF	value management framework

How to Use the VM Framework

The VM Framework has been structured to permit rapid interrogation. The first sections (executive summary and sections 1 to 3) describe the background and introduce the VM Framework.

Section 4 describes the pre-requisites characteristic of any value management workshop at any stage in the construction process and discusses the benefits and disadvantages and of any particular approach.

Section 5 describes the characteristic value management workshop in terms of an extended job plan highlighting the full VM process and the place of the workshop within that process.

Section 6 highlights the six value management opportunities identified through benchmarking as being the most common application points of value management during the project life cycle. This section introduces the following six sections that describe each workshop type under the headings described in section 5. Under each heading are listed the tools and techniques most commonly applicable at the particular stage in the process of the particular workshop type. The tools and techniques are referenced forward to the toolbox (section 16) which gives a short explanation.

Sections 13 and 14 discuss implementation and post-workshop issues and section 15 concludes.

As an example of a way to use the manual, consider someone well versed in value management and wishing to compare current techniques with those used by other practitioners around the world and included in this framework. The reader would, through the contents list, select an appropriate workshop type, for example a briefing workshop as described in section 8. Section 8 commences with diagrams illustrating the application point and typical workshop structures. There is then a list of commonly used tools and techniques listed under headings. The reader may wish to refer to:

- section 5 for a description of workshop stages
- section 6 to check on the definition of the briefing stage
- section 16 for a description of a particular techniques.

It is stressed now and later that this framework is not a textbook but a benchmark of international value management as practised in 1997/8.

Executive Summary

Value management (VM) is defined here as a proactive, creative, problem-solving or problem-seeking service which maximises the functional value of a project by managing its development from concept to use. The process uses structured, team-oriented exercises that make explicit and appraise existing or generated solutions to a problem, by reference to the value requirements of the client.

An alternative definition is that VM is concerned with making explicit and exploiting the package of benefits a client requires at an appropriate cost.

The research for this value management framework (VMF) document was conducted by the School of Civil Engineering at the University of Leeds and the Department of Building Engineering and Surveying at Heriot-Watt University, Edinburgh, as part of an EPSRC IMI-funded programme.[1]

The VMF has been developed from an international benchmarking exercise of value management processes, procedures, tools and techniques used in construction and manufacturing. The initial benchmark datum chosen for the benchmarking exercise was the value management process developed and implemented in the UK construction industry by Kelly and Male (1993). It should be noted here that the Kelly and Male value management process was used only as a frame of reference for the benchmarking partners in order to aid them in making comparisons and contrasts against their own practice of VM.

The VMF presents a framework for undertaking the VM process on projects, primarily in the construction industry. However, as experience has also been drawn from the manufacturing sector, large portions of this VMF may also be applicable to this industry. To gain a better understanding of the reasons for adopting the framework presented in this VMF, reference should be made to the research report.*

The VMF has been developed to be generic in nature and, as such, individual project procuring clients and VM practitioners must adapt the framework presented to meet their specific organisational and project needs. The VMF contains the important steps in the VM process and also the requirements at specific value opportunity points in the project life cycle. Clients and practitioners can use this framework to develop specific agendas and procedures to ensure that they will receive the maximum benefit from the tools and techniques presented.

Essential pre-requisites to successful VM are highlighted and discussed in section 4. The generic VM process (Figure 1) is illustrated and the steps outlined. Six value opportunity points (Figures 2 and 3) in the project life cycle, together with their associated tools, techniques, timings, durations and participants, are described in detail.

---

*The value management benchmark: Research results of an international benchmarking study is a companion document which contains details of the development of the framework, highlights important issues and draws conclusions. A CD-ROM containing an interactive version of the framework and the research report has also been produced. Both the research report and the CD-ROM are available from Thomas Telford Publishing.

[1] EPSRC IMI Programme. Construction as a Manufacturing Process.

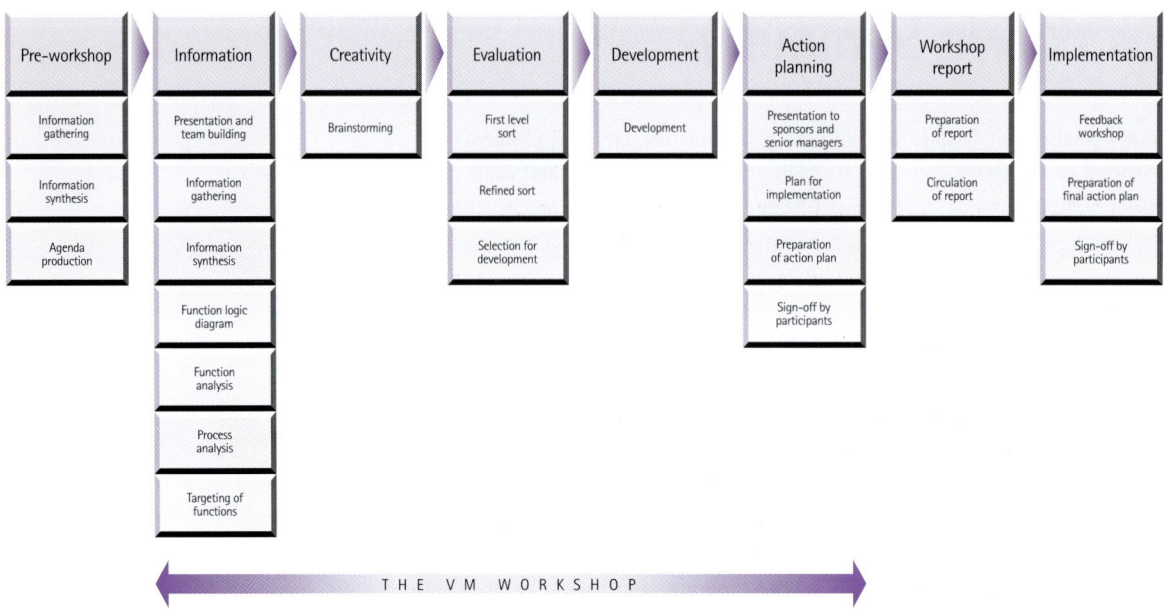

Figure 1 The generic VM process

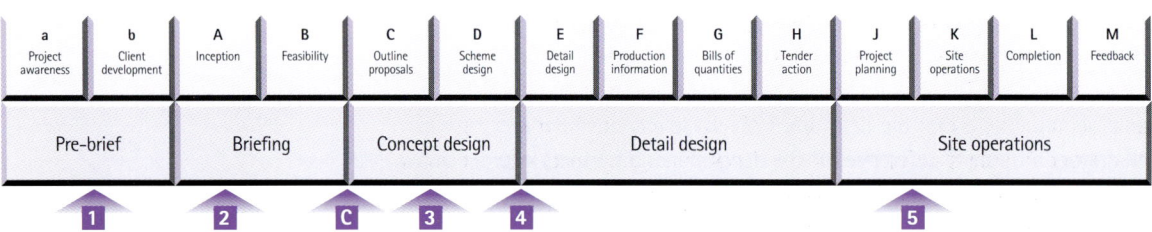

Figure 2 Value opportunities (RIBA)

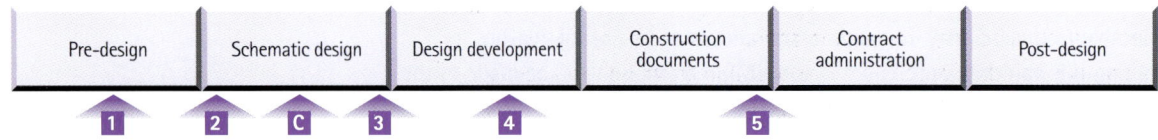

Figure 3 Value opportunities (AIA)

Benchmarking partners highlighted the following as being the critical success factors (CSFs) for VM:

- multi-disciplinary team / appropriate skill mix
- the skill of the facilitator
- the structured approach through the 'VM process'
- a degree of VM knowledge on the part of the participants
- presence of decision takers in the workshop
- participant ownership of the VM process output
- preparation prior to the VM workshop
- the use of functional analysis
- participant and senior management support for VM
- a plan for implementation of the workshop outcomes.

Some practitioners stated that VM workshops could not be successful unless all of the above CSFs were present.

1. Background to the Value Management Framework and Research Study Methodology

In the absence of a UK or international VM framework for construction the research objective was to undertake comparative research internationally with construction and manufacturing industries to improve project and business processes within client, design and construction team organisations. Figure 1.1 sets out the graphical representation of the research methodology.

Initially a review of literature was undertaken resulting in a synopsis of value management methodologies used internationally at various stages of the project/product life cycle in construction and manufacturing. An initial 'paper' benchmarking exercise was undertaken against the Kelly and Male (K&M) methodology to sensitise the research team to interview, case-study and benchmarking requirements. Interviews and case studies were undertaken with UK collaborators in construction and manufacturing to determine practice, procedures, methodologies and techniques used. This highlighted the strengths and weaknesses of various procedures at each stage of the project/product life cycle. Particular emphasis was given to client/customer viewpoints on the delivery and appropriateness of VM studies.

International fieldwork was undertaken in the UK, mainland Europe, North America, and Australasia. Construction and manufacturing were targeted and a method similar to the UK fieldwork was adopted. A comprehensive review of literature, rather than a series of interviews, was adopted for Japan.

Finally, prior to the launch of the VMF a series of regional workshops, comprising collaborators drawn from client, design and construction teams and professional associations, were held to appraise its contents critically. An opportunity was also given to all the benchmarking partners to appraise the contents of the VMF and make comments for consideration by the team.

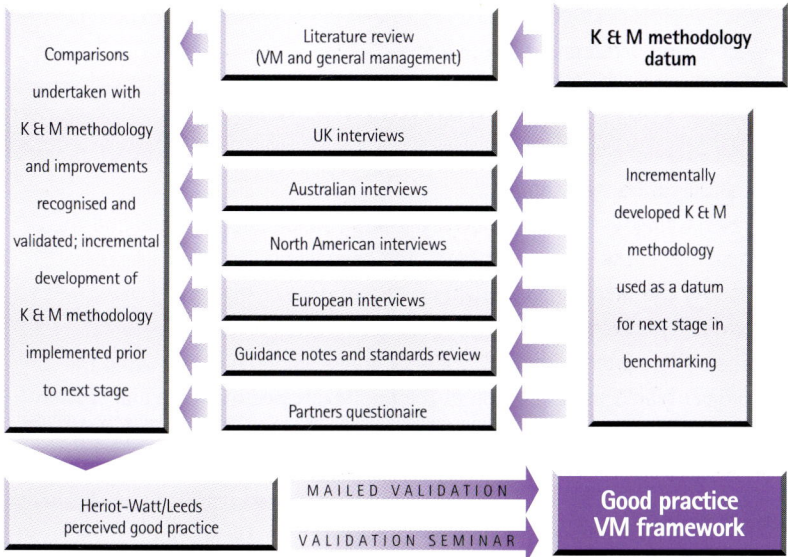

Figure 1.1 The research methodology adopted

2. A Guide to the Value Management Framework

The purpose of this document is to define and present a framework for the VM process and provide an outline for use on live projects. The framework should be considered the benchmark, a frame of reference based on current international practice of VM, to be used by practitioners and clients alike.

Unlike manufacturing projects dealing with the production of many identical products, construction projects (excluding housing) tend to involve the production of a single product. However, the underlying VM process remains relatively consistent across construction projects and this document presents VM at this generic level.

The VMF is primarily aimed at the UK construction market, specifically at clients who do or may commission VM studies and value management practitioners who will readily understand most of the terms and terminology in this document. Those unfamiliar with VM are recommended to look at one or more of the following guides to value management before reading this document.

- Building Research Establishment, *Value from Construction – Getting Started in Value Management,* BRE, 1997.
- Institution of Civil Engineers, *Creating Value in Engineering.* ICE Design and Practice Guide, Thomas Telford, 1996.
- Construction Industry Research and Information Association, *A Clients Guide to Value Management in Construction.* CIRIA, 1995.
- Building Services Research and Information Association, *Application Guide – Value Engineering of Building Services.* BSRIA
- H M Treasury, Central Unit on Procurement. *No. 54 Value Management.* HM Treasury, 1996.
- Construction Industry Board, *Fact sheet on – Value Management.* CIB, 1997.

Generic VM construction activity is targeted at five key value opportunities where the process could be employed to achieve maximum effect on any project during its life cycle. These are:

1. The pre-brief workshop – the identification of the client's value system and project mission, and the ensuring of its strategic fit with the corporate objectives of the client organisation. A decision to construct may be taken following this workshop.
2. The briefing workshop – undertaken after the decision to build and the appointment of the project team. The objective of the workshop is to specify the client's requirements in the context of the client's value system and the project mission.
3. The outline sketch design (OSD) workshop – undertaken following the project team's interpretation of the brief expressed in drawn and written form. This workshop should precede the application for detailed planning consent.
4. The final sketch design (FSD) workshop – undertaken once sketch design is complete but prior to the commencement of final sketch design. This workshop will audit the design against the brief and will address element and component issues.

5. The operations workshop – undertaken after the appointment of and including the contractor and key specialist sub-contractors. This workshop will address 'constructability' and supply chain issues.

In addition to the workshop styles above the widespread use of the 'Charette' is recognised (as 'C') on the value opportunity diagram. The Charette is a hybrid of the brief and the outline sketch design workshop and is used as a first value opportunity in situations where value management has not been used prior to the production of the brief and often the production of outline drawings.

Throughout the VMF comments made by VM consultants and clients have been placed in the margin of the text to reinforce and clarify points in the main body. The VMF is not intended to be a text book of value management; should greater detail be sought about the various tools and techniques, appropriate text books (listed in the references) will need to be consulted.

3. The Value Management Process

3.1 What is Value Management?

VM is defined here as a proactive, creative, problem-solving or problem-seeking service which maximises the functional value of a project by managing its development from concept to use. The process uses structured, team-oriented exercises that make explicit and appraise existing or generated solutions to a problem, by reference to the value requirements of the client.

An alternative definition is that VM is concerned with making explicit and exploiting the package of benefits a client requires at an appropriate cost.

A VM service is defined as one that involves:
- operating in multi-disciplinary or multi-expert teams
- using the VM process
- distilling information and analysing function or objectives prior to creativity
- establishing comparative cost in relation to function and worth and hence being overtly concerned with issues of value and value-for-money.

Value opportunities arise at points in the project life cycle when there is:
- an unstructured problem or opportunity
- a need for strategic commitment
- a convergence of information from different parties
- project uncertainty
- an introduction of new personnel to the project
- a need for technical commitment
- a need for capital commitment.

Figure 3.1 illustrates the use of the RIBA Plan of Work simply as a framework to illustrate value opportunities. It is a commonly used industry framework and should not be seen as an endorsement of any particular procurement route. It is very much seen as generic and can be understood by most construction professionals and clients. The American Institute of Architects (AIA) Design Process is also used here similarly (figure 3.2).

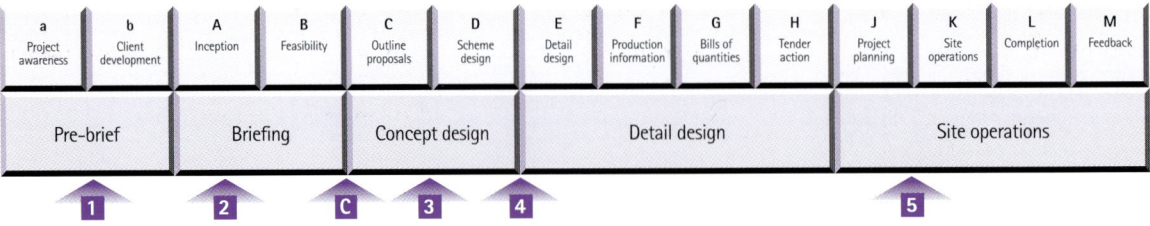

Figure 3.1 Value opportunities mapped out on a modified RIBA Plan of Work

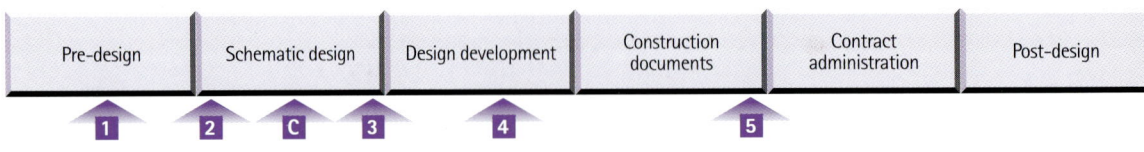

Figure 3.2 Value opportunities mapped out on an AIA Design Process

3.2   Why do Value Management?

Benchmarking collaborators stated,
'The advantages of VM include team building, shared knowledge and understanding, common focus on value structures and objectives, and flushing out areas of unnecessary cost.'

'VM is a good information exchange mechanism, produces a setting where valuable exchange of information and ideas can take place.'

'VM is a very good tool for breaking existing perceptions, to force people to take a fresh approach to problem solving and assisting in setting out tasks and objectives with value-for-money at the forefront of their thinking.'

VM forces people to address value, value-for-money and the benefits to be derived from a project, process, component or element, etc. and not just to focus on cost.

'VM techniques are far better than TQM techniques. TQM is how do we do it better? VM is why do we do it in the first place?'

'A VM study is often the first time the whole team has been in the same room at once.'

3.3   The Value Management Process

Figure 3.3 summarises the generic VM process distilled from the literature and benchmarking exercises in the UK, Europe, Australia and North America. The VM process is structured around the 'job plan' under eight headings. The VM workshop is only part of the overall process. These headings have been developed further than those in the more traditional texts. An explanation of the various terms can be found on subsequent pages.

The implementation of workshop suggestions is regarded as a separate but essential process that is carried out after a VM workshop has been completed. This is discussed further in section 10.

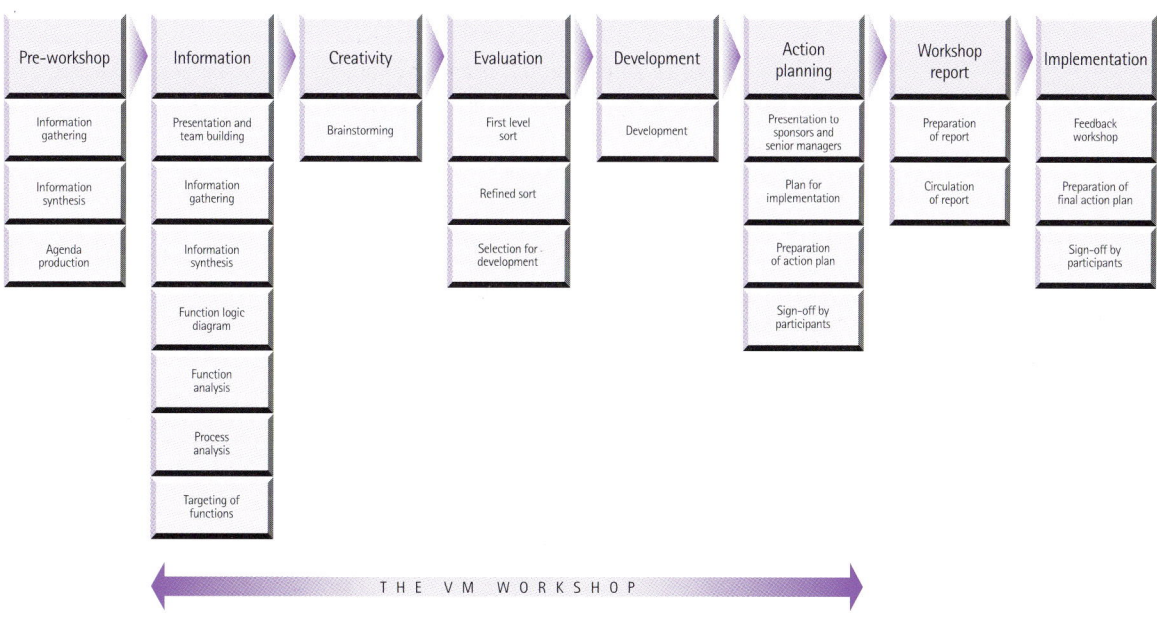

Figure 3.3 The generic value management process

4. The Pre-Requisites of the Value Management Process

This is an area that has not been given much consideration in the past. Pre-requisites are the conditions that must be satisfied to ensure a smoothly running value management process, mainly relating to the involvement of people and the venue for the workshop. These include:

- agreement to participate by all parties
- senior management support for value management
- experienced facilitator(s)
- appropriate team skill mix
- presence of client decision taker(s)
- isolated workshop environment
- commitment to the outcome.

4.1 Agreement to Participate by all Parties

'If you convince the team and client of the worth of VM then you will undoubtedly get good results.'

There is a need for the project team members to agree with any VM workshop requiring their participation, otherwise there is likely to be a 'blocking' atmosphere that can severely hamper the VM process.

4.2 Senior Management Support for VM

'If executive support is not overt a VM study will not even get off the ground.'

'It is important that the participants face the same direction.'

Senior management support for VM and its associated process is a pre-requisite of success. Senior management sets the scene by endorsing VM and giving it its backing to ensure that everyone knows that ideas and suggestions arising from workshops will be implemented. This signals to participants that what they say and do will have an impact on the project.

4.3 Experienced Facilitator(s)

'There needs to be a professional approach from the facilitator, i.e. people management, otherwise the team will not buy in.'

'Facilitators that are flexible, adaptable and able to judge situations radically improve studies.'

'The facilitator should be sensitive to team members' personalities as well as their professional skills.'

'Facilitation is a skill that should not be taken for granted. Facilitation is advisory, not driving.'

Experienced facilitators are of paramount importance for a successful VM workshop. The facilitator is, in effect, a VM process manager and acts as team co-ordinator, guiding the team through the tools and techniques best matched to the stage of project under investigation. Key skills include:

- setting the workshop agenda
- managing the process of the workshop, including:
  - dealing with any hidden agendas of participants
  - recognising individuals and their contribution
  - questioning and summarising
  - providing direction and a sense of common purpose
  - sensing interpersonal relationships within the team
  - sensing the climate of the workshop
  - synthesising and integrating information during the workshop
  - intervening in the workshop process when necessary
- communicating verbally and producing the VM report.

4.3.1 External vs. Internal Facilitator

An external facilitator is defined as someone who is neither explicitly involved with the project being analysed nor has a hierarchical relationship with members of the team, superior or subservient. An external facilitator also has no vested interest or hidden agendas with regard to the project. In this respect, external facilitators are considered better than internal because they are more likely to pose probing and challenging questions without the fear of sounding superior or inanely knowledgeable.

4.3.2 Number of Facilitators

Dual facilitation (two facilitators as opposed to one facilitator with a recorder) has emerged as the preferred option due to the mutual help the facilitators can give each other and also the variety it provides to the workshop participants. Two facilitators alternating can keep the momentum going to a higher degree than someone working on their own and it is possible to work longer days without sacrificing quality.

'The facilitator is the key! Two works better than one, takes pressure off and keeps the interest of the group.'

It should be noted, however, that the economics of the market place tend to drive clients towards the single-facilitator option, or sometimes one facilitator with a recorder. Clients may, therefore, not be getting value-for-money from the VM process.

4.4 Appropriate Team Skill Mix

In the debate on whether to use a completely independent workshop team or the existing project team, the latter emerged as the preferred basis for selecting the VM workshop team, bringing in experts to the workshop if required. An independent team consists of an independent group of experts with no connection to the project whatsoever and is usually chosen by the facilitator. The advantages of using the existing team include:

'Selecting the right study team with credibility is very important to ensure the study is successful and to achieve maximum implementation.'

- costs are kept down
- time briefing the team about the project is reduced
- unnecessary exploration of previously generated and failed solutions can be avoided
- better ideas may crop up due to greater experience 'second' time around
- there is another chance to explore alternative options, some of which may have been prematurely discarded or were waiting an appropriate time for implementation
- VM workshops are useful for team building and resolving conflicts that may exist within the existing team
- development time is considerably reduced as the team members will accept ideas for development in a cruder state than if they were being 'sold' an idea by an independent party
- implementation is increased.

Participants, excluding the facilitator, could include:
- architect
- structural engineer
- services engineer
- project manager
- other designer
- quantity surveyor or cost engineer
- client (may be by multiple representation depending on the project type); client representation will be appropriate to the stage of the project development, i.e. board level at the strategic stage, operational staff at the briefing stage
- end user
- facilities manager
- process engineer
- expert or specialist
- any recognised stakeholder in the project.

4.5   Presence of Client Decision Taker(s)

'Client decision takers are very important to have in a study. In other words, people who understand the project management process and can make it happen afterwards.'

The term 'decision taker' is used to represent a person who has the authority to make and take decisions during the workshop, whereas a 'decision maker' can only suggest solutions and needs to refer to a higher authority in order for the solutions to be ratified and then implemented.

The empowerment of the participants of a VM workshop team to take decisions will influence the effectiveness of the workshop. This impacts on the choice of an existing or independent team for the workshop, as discussed in section 4.4 and illustrated in greater detail in figure 4.1.

	Existing project team employed	Independent team
	Potential vested interests Deep project knowledge Easier implementation	No vested interests Time for learning required Difficult implementation Greater technology transfer
Decision taker present	Very reliable output Maximum implementation	Reliable output Medium implementation
Decision maker present	Reliable output Medium implementation	Less reliable output Low implementation

Figure 4.1 The relationship between team composition and whether a decision taker is present or not

The role of the facilitator is influenced by the above relationships. With an existing team the facilitator has to manage the VM process whilst assuming a challenging role in order to assist the team members to 'break' from their traditional thinking on the project. On the other hand, for an independent team, the facilitator can be more oriented towards managing the VM process only.

It should be noted that the presence of a client decision taker may not be required for workshops conducted later in the project life cycle, as the client's value system should already be set in place. However, it would be necessary to bring one on-board if no VM workshops have been carried out and therefore the value system has not previously been made explicit.

4.6   Isolated Workshop Environment

The likelihood of a VM workshop being successful is enhanced if an isolated environment is provided for the participants. An area should be set aside where the workshop can be held without interruptions from outside agents. Advantages include:
- it focuses the team on 'the project'
- gestation occurs during the workshop process
- it commits the team
- it mitigates against partial attendance.

It is important to weigh up the cost of providing the isolated environment to maximise the outcomes versus the apparent cost. An isolated environment, for example in a hotel, does focus the team totally on the project under investigation, in a 'pressure cooker' environment, and it stimulates team building. The overwhelming weight of research evidence supports an isolated workshop environment.

4.7   Commitment to the Outcome

A value management workshop generally ends in a positive mood with the team being committed to the outcome. However, on return to their own offices and the consequential pressures caused by their absence at the workshop, individual members of the team can become less enthusiastic about implementing innovative suggestions. It is at this stage that 'blocking' can occur for no other reason than the team member finds it more expedient to tread the tried and tested route.

This post-workshop fall-off in commitment can be countered by concluding the workshop with a detailed action plan where specific tasks are distributed amongst team members with responsibilities assigned and a date for completion given. An implementation workshop is arranged for the earliest possible date following the main workshop. It is at the implementation workshop that team members report back to the client on the economic viability, technical feasibility and functional suitability of the workshop conclusions. This process is conducive to continuing enthusiasm for maximum value to the client.

5. The Value Management Workshop Process

5.1 Pre-Workshop

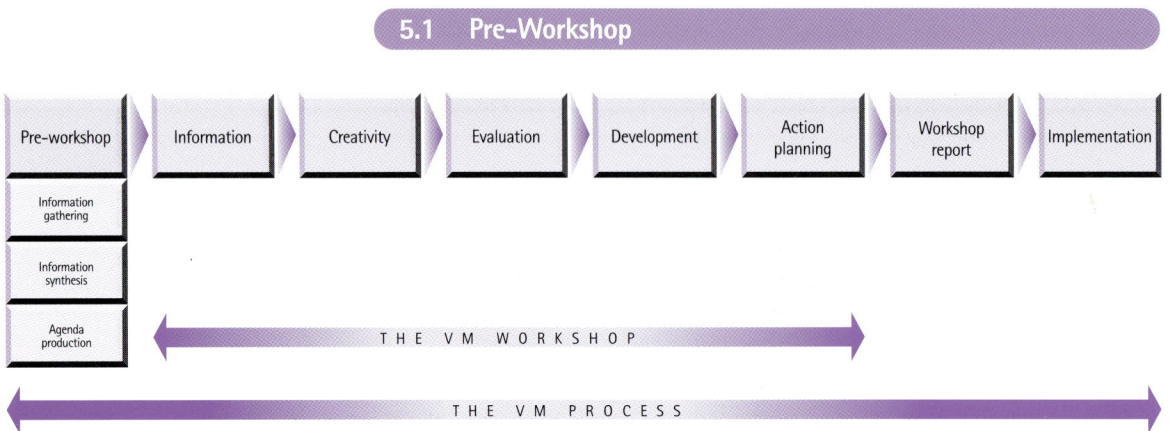

Figure 5.1 The pre-workshop phase

'The more time that is spent on pre-study information gathering and preparing the better the study will be.'

'Involve people before the study so that they are informed; make sure they understand why they are there.'

'If the pre-study information stage is conducted properly the facilitator will be well briefed regarding any hidden agendas and the politics behind the project.'

5.1.1 Information Gathering

Before a VM workshop commences information has to be gathered to determine its objectives and deliverables and therefore what shape and form the workshop will take. It is very important to 'get' to the client, to understand the client's thoughts and visions and to talk to various users to obtain their input. There is also a real need for the facilitator to recognise and establish any individual objectives and agendas of the participants.

The pre-study information gathering process does not alter, regardless of when in the project life cycle the workshop will be held, and all the tools and techniques detailed in the VM toolbox are valid.

The facilitator, the client or a combination of both, can undertake the choice of the VM workshop team. The important point is to ensure the right skill mix and personality of the participants. Typically those interviewed during the research indicated that they will spend two days to one week on pre-workshop information gathering, this being dependent upon project size, type, complexity and stage.

5.1.2 Information Synthesis – The Workshop Agenda/Information Pack

Once all the information has been gathered and synthesised the facilitator(s) produce the agenda for the workshop. The agenda should be accompanied by a summary of the pre-workshop information and circulated as an information pack to the workshop participants prior to the workshop. Alternatively, this synthesis process could be performed at a pre-workshop meeting of the workshop participants.

5.1.3 Number of Participants in a VM Workshop

There are various schools of thoughts on the ideal number of participants at a VM workshop. The number of people present should generally be between 5 to 15; it has been suggested that 5 to 7 is ideal (O'Donnell, 1994). However, this cannot be set in stone as one has to ensure that there is an appropriate set of skills and skill mix present to address the problem correctly. If there is a need for a large group, it is advisable to split them into smaller sub-groups that then come together at various points during the workshop to exchange information. Large VM teams may well occur at the front end of a project where a considerable number of stakeholders could be present. Large group workshops present a particularly challenging situation for the facilitator(s).

5.1.4 Physical Workshop Parameters

Physical workshop parameters that also need to be taken into consideration include the following.

- The layout of the room where the workshop will take place. A desk layout in the shape of a hollow square is considered ideal as it brings people relatively close together and increases eye contact between participants.
- The equipment that is required; standard office equipment should be provided with large areas for writing on and for pinning large sheets displaying suggestions and ideas. The most common medium for writing is the flip-chart. Other vital pieces of equipment include an overhead projector, sticky notes, pens, pencils etc. It should be noted that the use of electronic forms of visual presentation via notebook computers and computer projection is emerging as an important VM workshop technique.

5.2 VM Workshop Information Phase

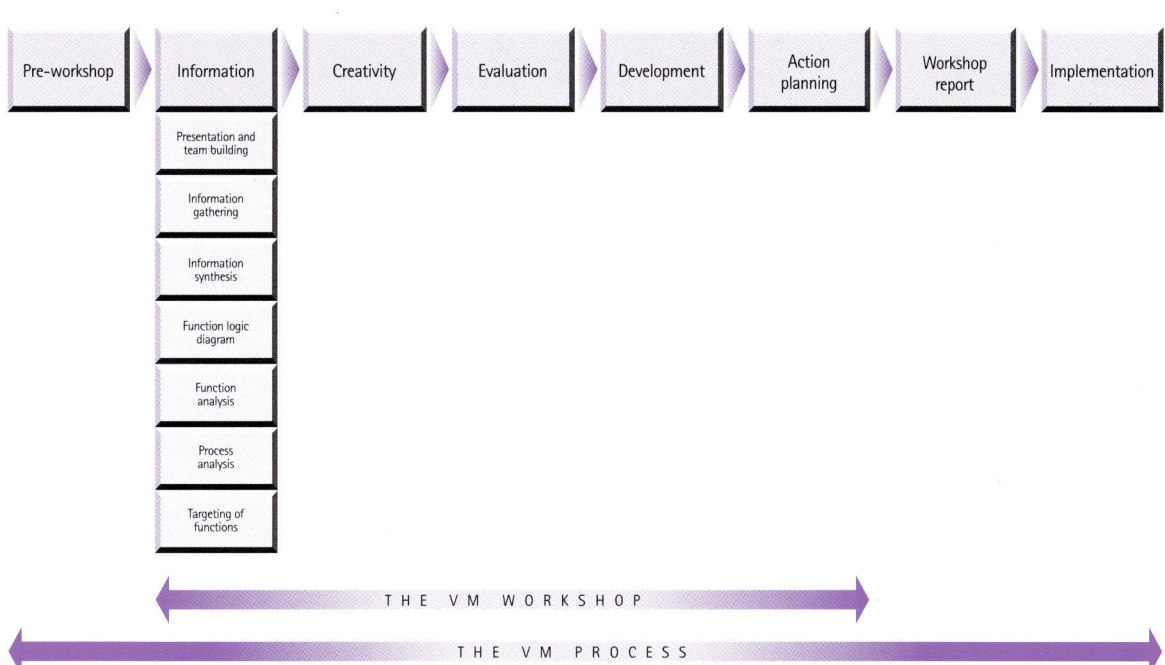

Figure 5.2 The VM workshop information phase

5.2.1 Information Gathering and Sharing

In this phase all of the available information relating to the project under review is gathered together. The objective of information sharing is to identify, in clear unambiguous terms, the issues and functions of the whole or parts of the project, as seen by the client organisation. The information should not be based upon assumption but obtained from the best possible source and corroborated, if possible, with tangible evidence and facts. The quality of any subsequent decision cannot rise above the quality of the information upon which it was based.

'The Information phase is crucial to binding the team together and getting them to buy-in to the VM process.'

The specific information sought at this stage is the following.
- Client needs. The fundamental requirements that the project must possess to serve the client's basic intentions. Needs should not be seen solely in

terms of utility. The client may have a need for a flamboyant statement or a need for a facility that heightens the client's esteem.

- Client wants. The embellishments that it would be nice to have but that do not satisfy need.
- Project constraints. Those factors that will impose a discipline, e.g. regulations, codes of practice.
- Budgetary limits. Expressed as the total amount that may be committed to the project in initial capital and life cycle costing terms.
- Time. For design and construction as well as the anticipated period over which the client will have an interest in the building.
- Quality, as expressed in client performance requirements.

This information forms the basis upon which the workshop team will conduct the function analysis (section 16.2.8).

Tools and techniques relevant to this part of the process are detailed in the toolbox (see section 16).

5.2.2 Regular Summaries
These should be undertaken at regular intervals throughout the workshop. They are important to ensure that a collective awareness is developing and no questions or issues remain outstanding.

5.3 VM Workshop Creativity Phase

Pre-workshop | Information | Creativity | Evaluation | Development | Action planning | Workshop report | Implementation

Brainstorming

THE VM WORKSHOP

THE VM PROCESS

Figure 5.3 The VM workshop creativity phase

5.3.1 Brainstorming
In the creative phase the VM team puts forward suggestions on solutions for the functions which have been selected for further investigation.

It is notable from benchmarking data that there is no variation of techniques applied at the creativity stage. Although the literature reveals there are a number of creativity techniques available (Kelly and Male, 1993; Norton and McElligott, 1995; Parker, 1985), brainstorming (section 16.3.1) proves to be by far the most popular and effective in practice.

5.4   VM Workshop Evaluation Phase

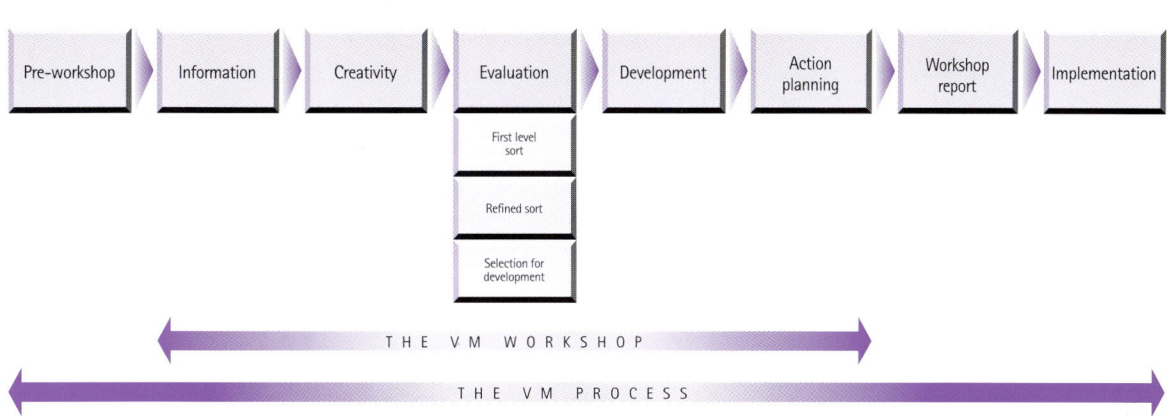

Figure 5.4 The VM workshop evaluation phase

The VM workshop team evaluates the ideas generated in the creativity phase. Several tools and techniques exist for evaluating ideas and suggestions that have been generated during the preceding creativity phase. Details of these can be found in the toolbox (section 16).

5.5   VM Workshop Development Phase

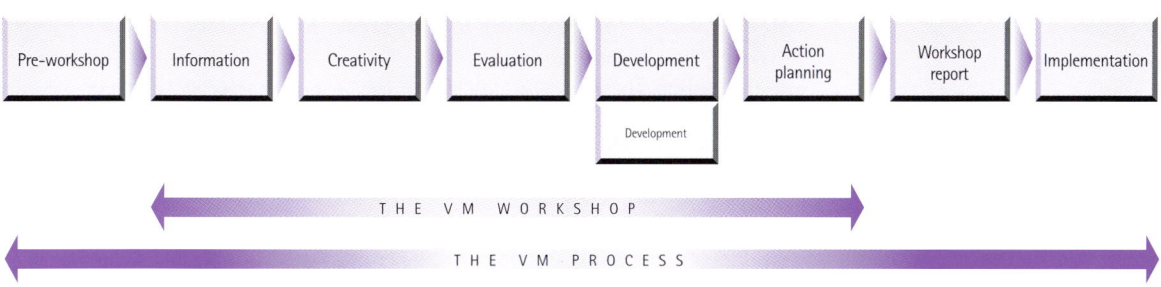

Figure 5.5 The VM workshop development phase

5.5.1   Development of Ideas

The accepted ideas, selected during the judgement phase, are investigated in considerable detail for their technical feasibility and economic viability. Outline designs will be worked out and costs determined. There is wide scope for the use of life cycle cost models and computer-aided calculations at this stage. It should be anticipated that only limited development can take place inside the workshop itself, further development should be carried out post-workshop but prior to an implementation review workshop. Details of further development should be detailed in the action plan and plan for implementation.

'What you need at the end of a formal study is a when, how and what is to be done statement that is then fed into the project process.'

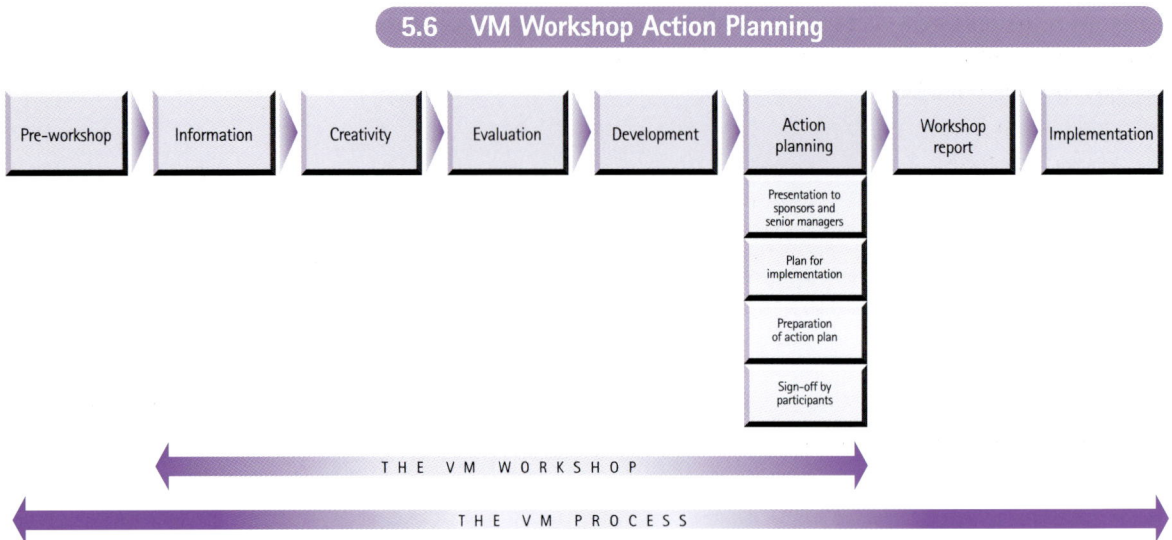

5.6   VM Workshop Action Planning

Figure **5.6** The VM workshop action planning phase

5.6.1   Presentation to Project Sponsors/Senior Managers

'A presentation is good as a closing technique, to focus the team and make sure everyone has a clear understanding of the next stage.'

The refined ideas are presented by the workshop team to the body commissioning the value management exercise. They are supported by drawings, calculations and costs, where appropriate. If a continuous workshop (where the various steps of the VM workshop process have been carried out in an uninterrupted sequence) has been held, then a presentation is a good way to end. It helps the participants focus on the main issues, ideas and solutions of the workshop. These presentations form the basis for the proposals to be implemented as a consequence of the workshop.

In the case of a workshop where sub-groups have been formed to develop ideas it is useful, as a matter of course, for each sub-group to present to the workshop members as a whole in a plenary session. This may or may not be in the presence of senior managers or project sponsors.

5.6.2   Plan for Implementation

The following planning is necessary:
- define the responsibilities of the workshop team members with regard to the proposals and work for further development of ideas
- define the deliverables of the proposals
- establish priorities
- establish a realistic timetable for implementation of workshop proposals
- establish a review process to check on the success of proposal implementation
- nominate an implementation manager.

5.6.3   Preparation of an Action Plan

The above checklist for the 'plan for implementation' is used as the basis for preparing an action plan to be incorporated in the final workshop report. This action plan functions as an audit to check on post-workshop progression of further development to ideas and proposal implementation.

5.6.4   Sign-off by Workshop Participants and Senior Management

To finalise a workshop, the workshop participants and senior management should sign a document detailing their agreement with the workshop findings,

their role in the action plan and their responsibilities for ensuring successful implementation of workshop proposals. Advantages of a sign-off include:
- greater implementation
- greater team focus
- greater project focus
- fewer disagreements further into the project life cycle
- accountability by all parties.

5.7 Workshop Report

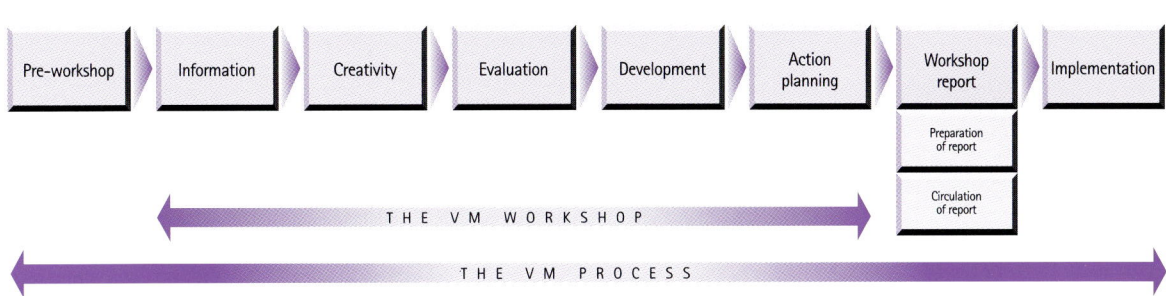

Figure 5.7 The VM workshop report phase

9.7.1 Preparation and Circulation of Report

A full report detailing the workshop process and findings, including the action plan, must be prepared as soon as possible after the workshop has ended. The report must then be circulated to the workshop participants to confirm their role in the implementation of the workshop proposals and any further development work necessary. The report will also aid the client in monitoring the proposals as they progress.

5.8 Summary

This section has outlined the VM process, highlighting each phase of the process and the important considerations needed at each phase. The next section sets out the possible VM value opportunities during the project life cycle.

6. Project Value Opportunities

6.1 Projects and Value Management

The previous sections dealt with the generic VM process. The focus will now be on how that process can be applied at various points in the project life cycle. In general terms VM aids in the following.

1. Establishing the relationship between the project and the client's corporate and business strategy and associated value criteria. Corporate strategy refers to the strategic direction of a number of companies grouped under one umbrella head office structure. Each of the companies could be operating in different markets or even in different industries. Business strategy refers to the strategic direction of a company competing in a particular market, market segment or industry.
2. Defining the project task or mission and asking the questions: why does the project exist? What problem does it have to solve? Is it in the right form? How does it relate to corporate or business strategy?
3. Establishing project objectives. What has the project to achieve? Are there conflicting objectives and if so, how should these be resolved?
4. Analysing the project functionality/time/cost/quality. Concerns include use and performance, fitness of purpose, choice of procurement route and contract strategy.

6.2 Value Management Workshops

'Once we value-managed a VM study and came up with a two-and-a-half day ideal. However, a lot of work has to be done before and after this type of study to make it work.'

Figure 6.1 illustrates the six value opportunities described in section 3.1. The value opportunity points are:
1. pre-brief workshop
2. briefing workshop
C Charette (undertaken in the place of the workshops 1, 2 and 3)
3. outline sketch design workshop
4. final sketch design workshop
5. operations workshop.

It should be noted that these workshops with their associated value opportunities should not be regarded as compulsory. They are simply the six most likely points in the project life cycle that the research has identified to be conducive to the use of the VM process. VM can take place at any point in the project life cycle and the number of value opportunities can and will vary from project to project. For example, a commissioning client organisation may decide to use VM at only one value opportunity point during the development of the project. Equally the client could decide to use three successive value management studies during the project.

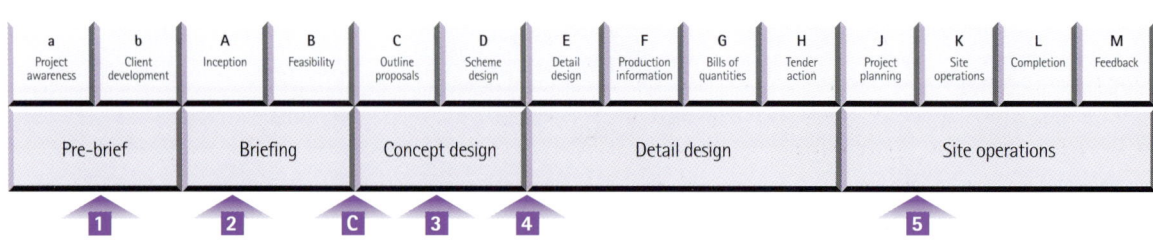

Figure 6.1 Value opportunities

6.3   Workshop Deliverables [2]

6.3.1   Pre–Brief Workshop – the Strategic Brief

The strategic brief sets out the broad scope and purpose of the project and its key parameters, including the overall budget and programme. It should provide an output specification that explains in clear terms what is expected of the project. One of the prime purposes of the pre-brief workshop is to present in clear and objective terms the mission of the project and its strategic fit with the corporate aims of the client organisation. Once this has been achieved a decision to build can be taken with confidence that all of the relevant issues have been addressed and alternatives examined.

'The VM process must try to keep the study duration to a minimum. To do this adequate time at pre-study is necessary and development is done outwith the study environment.'

The strategic brief output document should include:
- the mission statement
- the project context
- the client's value system and particularly how the success of the project will be measured
- organisational structure
- overall scope and purpose of the project
- programme, including phasing
- a global capital expenditure budget and cash-flow constraints
- targets and constraints on operating expenditure and other whole-life costs
- an implementation plan including the decision to build or factors to be considered in the decision to build.

6.3.2   Briefing Workshop – Project Brief

The project brief converts the strategic brief into construction terms, specifies the performance of the elements of the project and gives them an outline budget. The project brief output document should include:
- a summary of the relevant parts of the pre-brief document
- the aim of the design, including priorities for project objectives
- the functions and activities of the client
- the structure of the client organisation
- the site, including details of accessibility and planning
- the size and configuration of the facilities
- the project execution plan
- key targets for quality, time and cost, including milestones for decisions
- a method for assessing and managing risks and validating design proposals
- the procurement process
- environmental policy, including energy
- outline specifications of general and specific areas
- a budget for all elements
- options for environmental delivery and control
- servicing options and specification implications, e.g. security, deliveries, access, work place, etc.

6.3.3   Outline Sketch Design Workshop

Once the project sponsor has agreed the project brief, the project team can begin to test the design options that will contribute towards the concept design. For most projects the concept design is defined as that required in obtaining detailed planning permission. The outline sketch design workshop output document should include:

---

[2] Adapted from *Briefing the Team*, Construction Industry Board 1997

- a statement of the design direction
- the project execution plan
- the procurement plan
- key milestones
- performance measures
- the risks, a risk management strategy and detailed cost plan
- a detailed budget and schedule of activities
- the site layout and access, identifying ground conditions and planning constraints
- dimensioned outline drawings
- an outline specification for all systems.

10.3.4 Charette

A Charette is an audit of the brief often undertaken before the concept design is complete. In North America this workshop is referred to as being undertaken at 10% design. A Charette is commonly the first workshop on the project, which implies that the client has reached the decision to build, completed the brief and appointed the design team before undertaking a value management study. The study incorporates all the activities of the three workshops described above and is primarily used to check that the brief, and often the outline sketch design, comply with the client's value system. The first activity of the workshop is to overtly describe the value system of the client where this has not previously been expressed.

10.3.5 Final Sketch Design Workshop

Once the client project manager has 'signed-off' the outline sketch design, the project team should begin the development of the final sketch design and specification of the performance requirements for elements of the facility. The final sketch design should freeze as much of the design as possible, defining and detailing every component of the construction work. It should identify risks associated with the project and outline proposed action if they arise, assess the quality requirements and define how success will be measured. The detail design should include:

- a statement of scheme design
- the project execution plan
- key milestones and targets
- performance measures
- location of site, information on planning approvals and other detailed permissions agreed
- dimensions of spaces and elements provided
- performance specifications for environmental systems and services
- the cost plan
- proposals for the maintenance and management of the completed facility.

Each of the value opportunities highlighted above are detailed in the following sections and should not be considered to be prescriptive or rigid but indicative, flexible and able to be adapted to specific projects and approaches to VM.

6.3.6 Operations Workshop

The operations workshop is undertaken after the completion of the final sketch design and at the point at which construction work is about to commence. The operations workshop output should include:

- a statement of the extent of design consistent with the procurement route
- the project execution plan

- key milestones and targets
- performance measures
- a supply chain diagram
- pinch points or gates in project development, which have a strategic or tactical impact on following work packages
- identification of key work items to be targeted for specific technical workshops.

The exact definition of this stage will depend upon the procurement method adopted. The operations workshop will introduce supply chain and technical development issues. It should update the risks associated with the project and appraise the proposed action identified earlier.

6.4   Summary

This section has identified the value opportunity points. The next section will describe these in detail, including workshop duration and possible tools and techniques.

7. The Pre-Brief Workshop

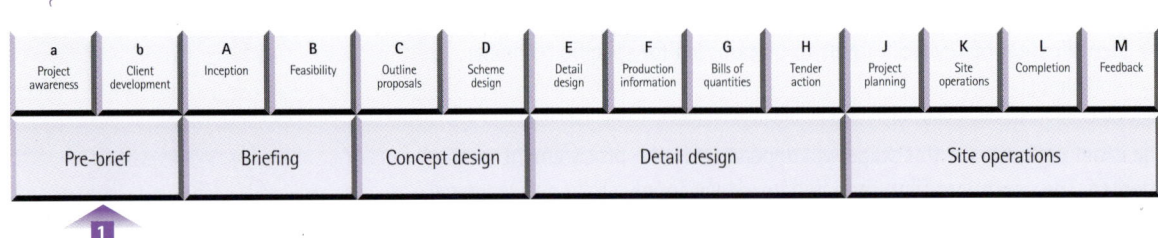

Figure 7.1 Pre-brief workshop related to project life cycle

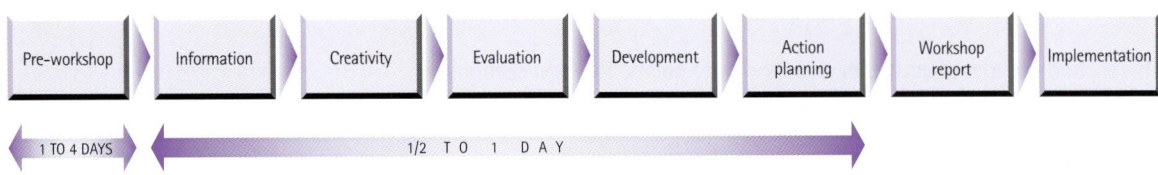

Figure 7.2 Pre-brief workshop duration and structure.

'We try to hit earlier in the project life cycle, around concept, as VM is more powerful at that stage.'

The primary aim of this workshop is to develop a strategic brief.

Indicative duration: ¹/₂ to 1 day
Potential participants: 10 to 20, all at senior level in the client organisation

7.1   Pre-Workshop Information Phase

Relevant techniques include:
- Checklist and interviews (sections 16.1.1 and 16.1.2)

7.2   VM Workshop Information Phase

Relevant techniques include:
- presentation and team building (section 16.2.1)
- brainstorm issues, group and theme (section 16.2.2)
- stakeholder analysis (section 16.2.3)
- strategic time line (section 16.2.4)
- project driver analysis (section 16.2.5)
- time/cost/quality analysis (section 16.2.7)
- function analysis (section 16.2.8)
- function logic diagram (section 16.2.9)

7.3   VM Workshop Creativity Phase

Relevant techniques include:
- brainstorming (section 16.3.1)

7.4   VM Workshop Evaluation Phase

Relevant techniques include:
- silence means 'no' (section 16.4.2)
- 'coloured dots' (section 16.4.3)
- championing (section 16.4.4)
- big issues (section 16.4.5)

7.5   VM Workshop Development Phase

Relevant techniques include:
- establishment of project mission and outline specification (section 16.5.1)

7.6   VM Action Planning Phase

Relevant techniques include:
- presentations (section 16.5.6)
- plan for implementation (section 16.6.1)
- preparation of action plan (section 16.6.2)
- sign off by participants (section 16.6.3)

7.7   VM Workshop Report

Relevant techniques include:
- preparation and circulation of report (sections 16.7.1 and 16.7.2)

7.8   Output from this Workshop

The outputs from the pre-brief workshop are:
- the statement of project mission (section 16.8.1)
- a decision to build or the issues on which a decision to build is to be taken (section 16.8.5)
- a project execution plan for the whole project and agreed activity to the next workshop (section 16.8.6)
- a workshop report as a record of proceedings (section 16.8.8)

8. The Briefing Workshop

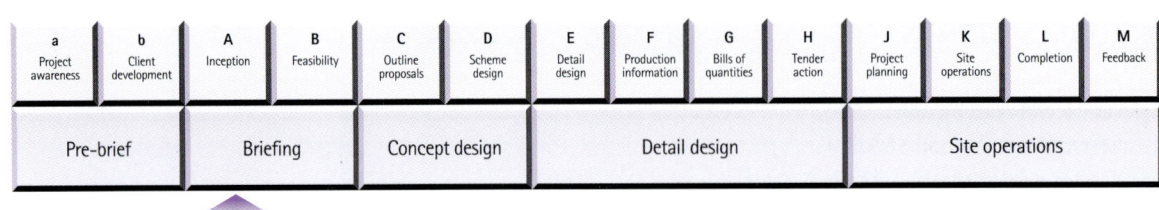

Figure 8.1 Briefing workshop related to project life cycle

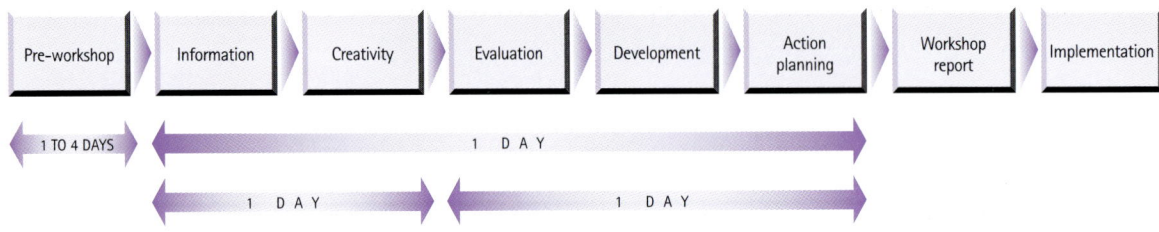

Figure 8.2 Briefing workshop duration and structure options

'Studies are generally shorter and less structured at the front end of a project. There is always a focus on function, however, which is seen as the key to good VM.'

The primary aim of this workshop is to develop the project brief.

Indicative duration: 1 to 2 days

Potential participants: 10 to 20, representatives of the client's managerial team, design and project management team

8.1   Pre-Workshop Information Phase

Relevant techniques include:
- checklist and interviews (sections 16.1.1 and 16.1.2)
- document analysis (section 16.1.3)
- questionnaire (section 16.1.4)
- post-occupancy evaluation (POE) of a similar facility or of the facility under discussion in the case of refurbishment and adaptation projects (sections 16.1.6)
- site tour (sections 16.1.5)
- facilities walk-through (section 16.1.7)

8.2   VM Workshop Information Phase

Relevant techniques include:
- presentation and team building (section 16.2.1)
- brainstorm, group and theme (section 16.2.2)
- stakeholder analysis (section 16.2.3)
- strategic time line (section 16.2.4)
- project driver analysis (section 16.2.5)
- time/cost/quality analysis (section 16.2.7)
- function analysis (section 16.2.8)
- function logic diagram (section 16.2.9)
- process flow charting (section 16.2.13)
- facilities walk-through (section 16.2.6)
- spatial adjacency analysis (section 16.2.10)
- REDReSS (section 16.2.11)
- SWOT (section 16.2.12)

8.3   VM Workshop Creativity Phase

Relevant techniques include:
- brainstorming (section 16.3.1)

8.4   VM Workshop Evaluation Phase

Relevant techniques include:
- silence means 'no' (section 16.4.2)
- 'coloured dots' (section 16.4.3)
- championing (section 16.4.4)
- big issues (section 16.4.5)
- technically feasible/economically viable/functionally acceptable/acceptable to client (section 16.4.1)

8.5   VM Workshop Development Phase

Relevant techniques include:
- detailed functional specification (section 16.5.2)

8.6   VM Action Planning Phase

Relevant techniques include:
- presentations (section 16.5.6)
- plan for implementation (section 16.6.1)
- preparation of action plan (section 16.6.2)
- sign-off by participants (section 16.6.3)

8.7   VM Workshop Report

Relevant techniques include:
- preparation and circulation of report (sections 16.7.1 and 16.7.2)

8.8   Output from this Workshop

The outputs from the briefing workshop are:
- a performance specification for the project including a detailed description of the project's component parts (section 16.8.2)
- an updated PEP for the whole project and agreed activity to the next workshop (section 16.8.6)
- a workshop report as a record of proceedings (section 16.8.8)

9. The Charette

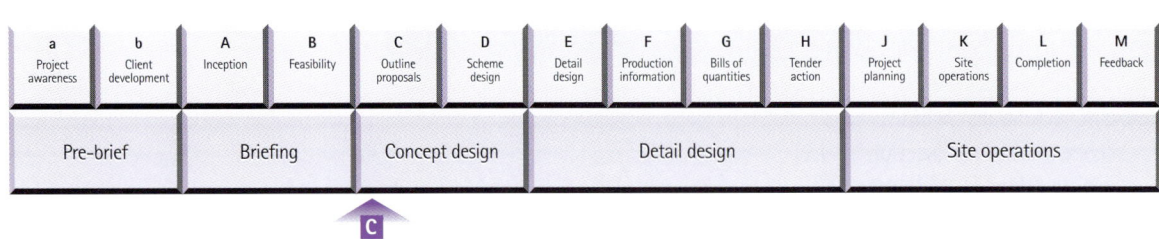

Figure 9.1 The Charette related to project life cycle

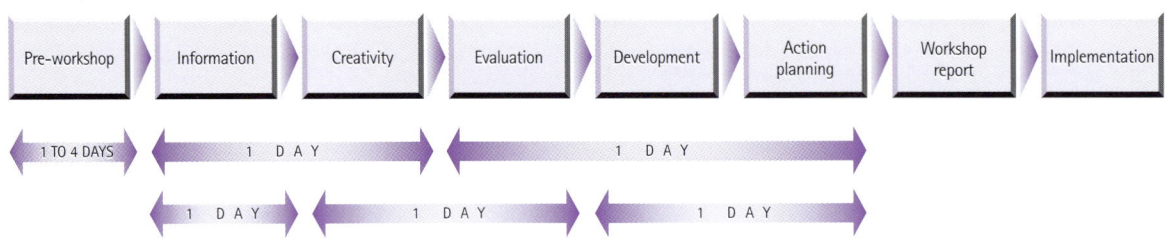

Figure 9.2 The Charette duration and structure options

'It often takes more than one day for people to change their views, therefore, two to three days' duration is seen as good. It gives enough time to fill the mind with knowledge, then to go through the process and generate solutions. The two-day working model is good, the process will then work and protect.'

The primary aim of this workshop is to review the brief and improve the concept design.

Indicative duration: 2 to 3 days
Potential participants: 10 to 15, including the design team, project management and client representatives

This workshop is appropriate where pre-brief and briefing workshops have not previously been held. A Charette may be undertaken immediately after the preparation of the brief or, more commonly, after the preparation of the outline sketch design. It is important, however, that the Charette is held before detailed planning consent is applied for, so that any necessary changes to the design can be undertaken without prejudice to the overall programmed time.

9.1 Pre-Workshop Information Phase

Relevant techniques include:
- checklist and interviews (sections 16.1.1 and 16.1.2)
- document analysis (section 16.1.3)
- site tour (section 16.1.5)
- questionnaire (section 16.1.4)
- POE reports (section 16.1.6)
- facilities walk-through (section 16.1.7)

9.2 VM Workshop Information Phase

Relevant techniques include:
- presentation and team building (section 16.2.1)
- brainstorm issues, group and theme (section 16.2.2)
- stakeholder analysis (section 16.2.3)
- strategic time line (section 16.2.4)
- project driver analysis (section 16.2.5)

- time/cost/quality analysis (section 16.2.7)
- function analysis (section 16.2.8)
- function logic diagram (section 16.2.9)
- process flow charting (section 16.2.13)
- facilities walk-through (section 16.2.6)
- spatial adjacency analysis (section 16.2.10)
- REDReSS (section 16.2.11)
- SWOT (section 16.2.12)

9.3   VM Workshop Creativity Phase

Relevant techniques include:
- brainstorming (section 16.3.1)

9.4   VM Workshop Evaluation Phase

Relevant techniques include:
- silence means 'no' (section 16.4.2)
- 'coloured dots' (section 16.4.3)
- championing (section 16.4.4)
- big issues (section 16.4.5)
- technically feasible/economically viable/functionally acceptable/acceptable to client (section 16.4.1)
- SWOT analysis (section 16.2.12)

9.5   VM Workshop Development Phase

Relevant techniques include:
- detailed functional specification (section 16.5.2)
- life cycle costing (LCC) (section 16.5.5)

9.6   VM Action Planning Phase

Relevant techniques include:
- presentations (section 16.5.6)
- plan for implementation (section 16.6.1)
- preparation of action plan (section 16.6.2)
- sign-off by participants (section 16.6.3)

9.7   VM Workshop Report

Relevant techniques include:
- preparation and circulation of report (sections 16.7.1 and 16.7.2)

9.8   Output from this Workshop

The outputs from the Charette are:
- a validated brief and concept design (section 16.8.2 and 16.8.3)
- an updated PEP for the whole project and agreed activity to the next workshop (section 16.8.6)
- a workshop report as a record of proceedings (section 16.8.8)

10. The Outline Sketch Design (OSD) Workshop

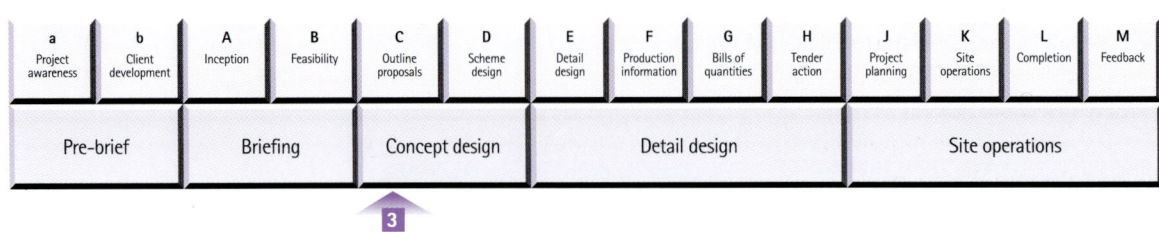

Figure 10.1 OSD workshop related to project life cycle

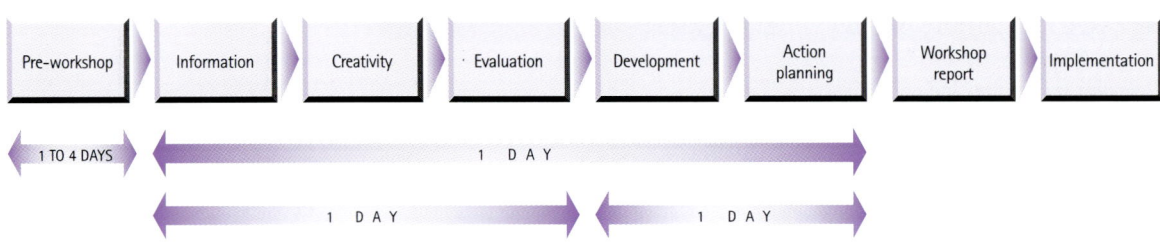

Figure 10.2 OSD workshop duration and structure options

The primary aim of this workshop is to review the brief and improve the concept design.

Indicative duration: 1 to 2 days
Potential participants: 10 to 15, including the design team, project management and client representatives at managerial and operations levels

This workshop is appropriate where pre-brief and briefing workshops have previously been undertaken. It is important that the outline sketch design workshop is held before detailed planning consent is applied for, so that any necessary changes to the design can be undertaken without prejudice to the overall programmed time.

10.1 Pre-Workshop Information Phase

Relevant techniques include:
* checklist and interviews (sections 16.1.1 and 16.1.2)
* document analysis (section 16.1.3)
* site tour (section 16.1.5)
* questionnaire (section 16.1.4)
* POE reports (section 16.1.6)
* facilities walk-through (section 16.1.7)

In a situation where a previous value opportunity workshop has been conducted, it may only be necessary to undertake the following:
* interviews as necessary, should any members of the team change (section 16.1.2)
* review of action plans of previous workshops (section 16.1.3)
* review of brief (section 16.1.3)
* assembly of drawings (section 16.1.3)

10.2   VM Workshop Information Phase

Relevant techniques include:
- presentation and team building (section 16.2.1)
- brainstorm issues, group and theme (section 16.2.2)
- stakeholder analysis (section 16.2.3)
- strategic time line (section 16.2.4)
- project driver analysis (section 16.2.5)
- time/cost/quality analysis (section 16.2.7)
- function analysis (section 16.2.8)
- function logic diagram (section 16.2.9)
- facilities walk-through (section 16.2.6)
- process flow charting (section 16.2.13)
- spatial adjacency analysis (section 16.2.10)
- REDReSS (section 16.2.11)
- SWOT (section 16.2.12)

In a situation where a previous value opportunity workshop has been conducted it may only be necessary to undertake the following:
- presentation and team building (section 16.2.1)
- review functional data
- review functional space and adjacency
- review drawings
- identification of mismatches

10.3   VM Workshop Creativity Phase

Relevant techniques include:
- brainstorming (section 16.3.1)

10.4   VM Workshop Evaluation Phase

Relevant techniques include:
- silence means 'no' (section 16.4.2)
- 'coloured dots' (section 16.4.3)
- championing (section 16.4.4)
- big issues (section 16.4.5)
- technically feasible/economically viable/functionally acceptable/acceptable to client (section 16.4.1)
- SWOT analysis (section 16.2.12)

10.5   VM Workshop Development Phase

Relevant techniques include:
- outline time/cost/quality analysis (section 16.5.3)
- LCC (section 16.5.5)

10.6   VM Action Planning Phase

Relevant techniques include:
- preparation and making of presentations (section 16.5.6)
- plan for implementation (section 16.6.1)
- preparation of action plan (section 16.6.2)
- sign-off by participants (section 16.6.3)

10.7 VM Workshop Report

Relevant techniques include:
- preparation and circulation of report (sections 16.7.1 and 16.7.2)

10.8 Output from this Workshop

The outputs from the OSD workshop are:
- a validated concept design (section 16.8.3)
- an updated PEP for the whole project and agreed activity to the next workshop (section 16.8.6)
- a workshop report as a record of proceedings (section 16.8.8)

11. The Final Sketch Design (FSD) Workshop

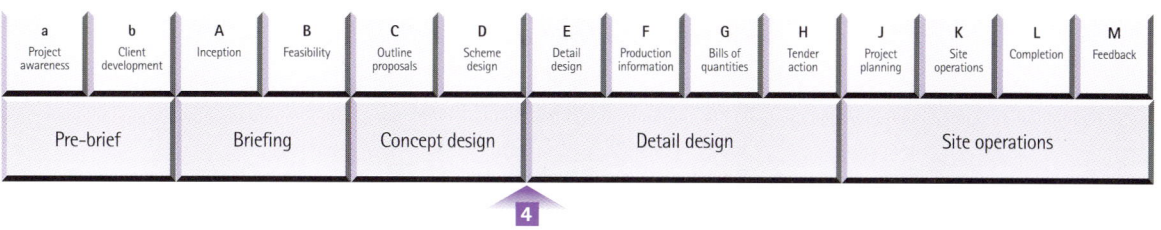

Figure 11.1 FSD workshop related to project life cycle

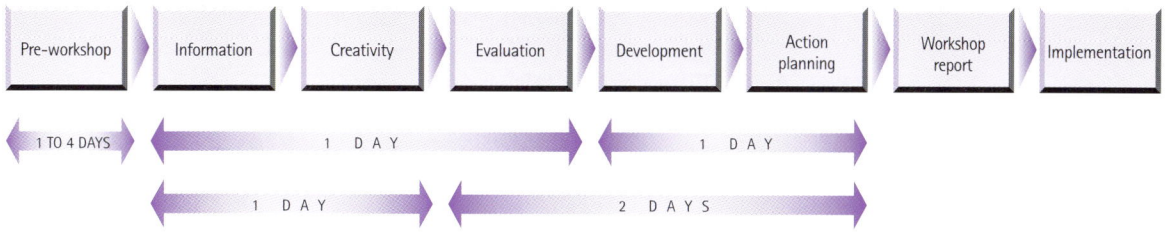

Figure 11.2 FSD workshop duration and structure options

The primary aim of this workshop is to fine-tune the design through element function analysis.

Indicative duration: 2 to 3 days
Potential participants: 10 to 15, including the design and project management team and facilities manager

This process assists the functional performance of elements through the correct choice of components to meet the function over the elements' life cycle.

11.1 Pre-Workshop Information Phase

Relevant techniques include:
- checklist and interviews (sections 16.1.1 and 16.1.2)
- document analysis (section 16.1.3)
- site tour (section 16.1.5)
- questionnaire (section 16.1.4)
- POE reports (section 16.1.6)
- facilities walk-through (section 16.1.7)

In a situation where a previous value opportunity workshop has been conducted, it may only be necessary to undertake the following:
- interviews (section 16.1.2)
- review of action plans of previous workshops (section 16.1.3)
- review of OSD (section 16.1.3)
- assembly of drawings (section 16.1.3)
- ascertainment of cost plan and preparation of histograms (section 16.2.15)
- identification of major work packages (section 16.1.9)

11.2 VM Workshop Information Phase

Relevant techniques include:
- presentation and team building (section 16.2.1)
- brainstorm issues, group and theme (section 16.2.2)
- stakeholder analysis (section 16.2.3)
- strategic time line (section 16.2.4)
- project driver analysis (section 16.2.5)
- time/cost/quality analysis (section 16.2.7)
- function analysis (section 16.2.8)
- function logic diagram (section 16.2.9)
- process flowcharting (section 16.2.13)
- facilities walk-through (section 16.2.6)
- spatial adjacency analysis (section 16.2.10)
- REDReSS (section 16.2.11)
- SWOT (section 16.2.12)

In a situation where a previous value opportunity workshop has been conducted, it may only be necessary to undertake the following:
- presentation and team building (section 16.2.1)
- strategic time line (section 16.2.4)
- element function analysis (section 16.2.16)
- component analysis (section 16.2.17)
- Pareto analysis – histogram of cost (section 16.2.15)
- value vs. cost (section 16.2.14)
- identify mismatches

11.3 VM Workshop Creativity Phase

Relevant techniques include:
- brainstorming (section 16.3.1)

11.4 VM Workshop Evaluation Phase

Relevant techniques include:
- silence means 'no' (section 16.4.2)
- technically feasible/economically viable/functionally acceptable/acceptable to client (section 16.4.1)
- decision matrices (section 16.4.6)

11.5 VM Workshop Development Phase

Relevant techniques include:
- detailed time/cost/quality analysis (section 16.5.4)
- LCC (section 16.5.5)

11.6 VM Action Planning Phase

Relevant techniques include:
- preparation and making of presentations (section 16.5.6)
- plan for implementation (section 16.6.1)
- preparation of action plan (section 16.6.2)
- sign-off by participants (section 16.6.3)

11.7 VM Workshop Report

Relevant techniques include:
- preparation and circulation of report (sections 16.7.1 and 16.7.2)

11.8 Output from this Workshop

The outputs from the FSD workshop are:
- a validated final sketch design (section 16.8.4)
- a PEP for the whole project and agreed activity to tender stage (section 16.8.6)
- a workshop report as a record of proceedings (section 16.8.8)

12. The Operations Workshop[3]

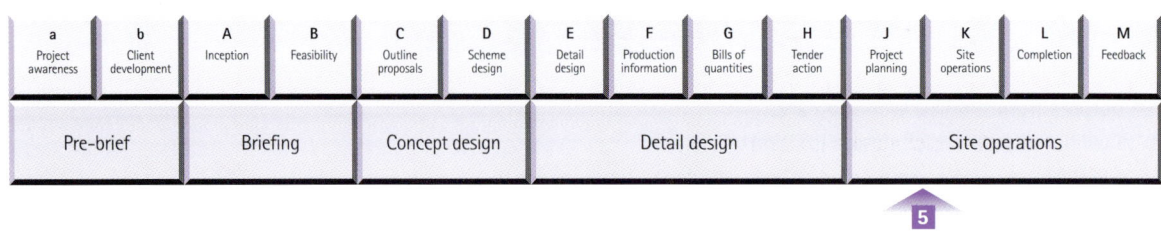

Figure 12.1 Operations workshop related to project life cycle

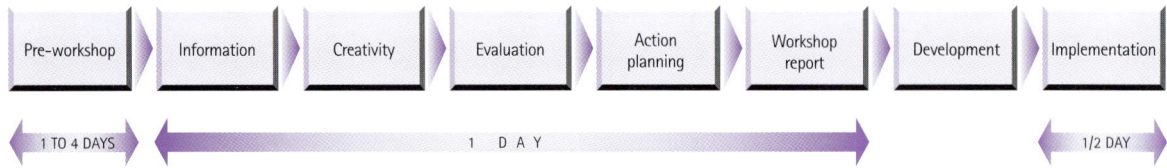

Figure 12.2 Operations workshop duration

'Studies are not necessarily continuous which is seen as an advantage because it gives people time to think and gather more data.'

The primary aim of this workshop is to convert the design into components and construction operational sequences.

Indicative duration: 1 or more days

Potential participants: 6 to 10 including production planning, purchasing, project management, supplier and/or sub-contractor

It is recognised that, of all the workshops described, this type is the most variable in terms of its structure – the focus could be the whole project, an individual work package or a component.

12.1 Pre-Workshop Information Phase

Relevant techniques include:
- interviews (section 16.1.2)
- understanding of contractor's viewpoint and strategy (section 16.1.11)
- assembly of production drawings and contract programme (section 16.1.8)
- preparation of histograms from tender information (section 16.2.15)
- identify major work packages (section 16.1.9)
- make explicit change management procedures (section 16.1.10)

12.2 VM Workshop Information Phase

Relevant techniques include:
- presentation and team building (section 16.2.1)
- review of contract programme
- supply chain analysis (section 16.2.18)
- functional specification (section 16.2.19)
- component analysis (section 16.2.17)
- Pareto analysis – histogram of cost (section 16.2.15)
- pinch point and committal point analysis (section 16.2.20)

---

[3] This section has been adapted from work undertaken by Nigel Standing, Southern Water, with respect to contractors' and site operations benchmarked against the value analysis procedures undertaken by manufacturers interviewed as part of the research.

12.3 VM Workshop Creativity Phase

Relevant techniques include:
- brainstorming (section 16.3.1)

12.4 VM Workshop Evaluation Phase

Relevant techniques include:
- silence means 'no' (section 16.4.2)
- technically feasible/economically viable/functionally acceptable/acceptable to client (section 16.4.1)
- decision matrices (section 16.4.6)

12.5 VM Action Planning Phase

Relevant techniques include:
- presentations (section 16.5.6)
- plan for implementation (section 16.6.1)
- preparation of action plan (section 16.6.2)
- sign-off by participants (section 16.6.3)

12.6 VM Workshop Report

Relevant techniques include:
- preparation and circulation of report (sections 16.7.1 and 16.7.2)

12.7 VM Workshop Development Phase

This phase of the workshop may be undertaken outside the formal workshop. Relevant techniques include:
- detailed time/cost/quality analysis (section 16.5.4)
- LCC (section 16.5.5)
- preparation of production plans

12.8 VM Implementation Workshop

Where the development phase takes place outside the workshop, this phase brings the team together to re-visit the action plan and draw up another.

12.9 Output from this Workshop

The outputs from the final sketch design workshop are:
- a validated production programme (section 16.8.7)
- a workshop report as a record of proceedings (section 16.8.8)

13. Implementation Workshop

The implementation workshop can be held after any previous workshop stage to allow members of the team to report on their elements of the action plan. Any mismatches in the reported outcomes are highlighted and solved. The primary aim of this workshop is to ensure that recommendations set out in the workshop report are being implemented.

Indicative duration: $^1/_2$ to 1 day
Potential participants: as in the workshop being followed up plus senior management

The scope of the workshop should include:
- recommendations and findings of the workshop being followed up and summarised in a presentation by the participants
- presentations made by workshop participants on progress on items detailed in the workshop action plan; implementation issues are included together with problems encountered
- brainstorming solutions to problems encountered or recommendations that have run into difficulties; these are reviewed, refined and prioritised
- brainstorming of new implementation issues, together with reviewing, refining and prioritising
- development of recommendations and an action plan
- a sign-off implementation workshop report by all participants.

14. Post-Workshop Issues

The major output from a value management workshop is the detailed report which will contain all the ideas generated that have been worked up and put forward. The report will form the basis on which ideas are put forward for implementation.

Proposals are developed in greater detail with action plans drawn up for their implementation or explanations produced as to why it is not possible to pursue them any further. It has been found that the implementation rate increases if:

- facilitators and design team have been working with the client from the inception of the project
- the existing design team is used in workshops. However, outside experts may sometimes be brought in
- there is a system to monitor progress, i.e. there is feedback loop in place
- ownership of ideas is ensured.

Implementation needs to be monitored and there is a need to have VM steering groups in companies, containing key stakeholders. They need to:

- agree on a VM programme
- refine resources that are to be allocated to the VM process
- monitor and ensure implementation.

'The report should serve as a vehicle for the decision makers on the project and inform people who are not directly involved on the project.'

'Need pressure to make people deliver suggested ideas; ownership of ideas is the only way of ensuring implementation.'

'Review session regarded as something that should be done to ensure implementation.'

15. Conclusions

'A VM study gets people to talk about the work they do, makes them question what they do. The tools and techniques are just a vehicle to achieve this.

'How well the people understand the process is very important in order for people to expand their minds.'

'There has to be a desire for change.'

'It must be recognised that there are some stress on the participants in a VM study.'

'The main advantage of VM is that it is the only way of ensuring quality AND competitiveness.'

The information stage is often cited as the most important stage in the VM process, laying the foundations on which the remainder of the exercise is built. The VM process diagram (figure 3.3) shows that the information stage has the greatest number of possible activities associated with it, particularly if pre-workshop information is also included.

Although figure 3.3 shows the process as a sequence of well-defined steps, the practice of VM is not so clear-cut. There is likely to be some overlapping and iteration of the team's activities.

The primary reason for any client commissioning a VM study is to achieve value improvements and/or monetary savings on the project under investigation to ensure value-for-money. This will be the case in the vast majority of VM workshops. However, there are also several additional benefits that will come from properly conducted VM exercises. These include the following facts.

- VM workshops are extremely useful for team building. It is to the benefit of the client if an existing team is used in the exercise. It will have the maximum benefit if it is in the early life of the project or early in the design stage when the team are 'coming together'.
- The workshop provides an environment for accelerating the gelling of the team and mutual understanding. It forces the participants to open up and commit themselves to the process of value management and also to the project itself.
- VM workshops promote client peace of mind. Even if a workshop may not yield great monetary savings or value improvements, the client can go away knowing that everything is proceeding according to plan and that there are no surprises looming on the horizon.
- There is also a tendency in workshops to allocate responsibilities to a greater degree than is usually the norm. People are made accountable for various parts of the project under investigation. This will have a positive impact as the project progresses.

An additional major benefit to the client of a value management workshop is that the process is, at least partially, an initial risk analysis of the problem under investigation. The tools and techniques outlined above lend themselves towards investigating problems critically and risks and perceived risks are consequently brought to the surface and highlighted.

At the end of a value management study the client will not only have reaped the traditional benefits of value improvements and monetary savings, but will also have gained an increased awareness of the risks involved together with team building. An allocation of responsibilities will have taken place. Finally, the client will feel more secure about where the project is currently, and where it is heading.

16. Value Management Toolbox

16.1 Pre-Workshop Information Phase Tools

16.1.1 Checklist

This technique is adapted from Morris and Hough (1987) and has been found to be generic across projects. It facilitates the extraction of necessary information and interactions, and also challenges existing perceptions and highlights to the facilitator any hidden agendas.

The checklist consists of the following headings:

Project environment – an exploration of the environment within which the project is to exist will sensitise the value management team to events that may unfold as the project is set in motion. Typical issues here would be major external environmental changes that can be foreseen which may impact the project development, for example:

- the timing of changes in government, both national, regional or local
- possible shifts in interest or exchange rates over the project life cycle
- the competitive impact of the industry or sector within which the project and facility are to operate
- demographic shifts.

Community – are there any community interests in the project, for example local interest groups?

Politics – the political parties in power at national, regional and local levels, their views on the project: are there any changes in political persuasion anticipated over the project life cycle and do these matter? Client organisational politics – the powerful departments, the powerful individuals or groups; who is to be represented on the client's project team, from which departments and how much power do they have to make decisions and influence project development?

Finance – how is the project to be financed, are there different sources of finance and is there likely to be any conflict? What is the client's funding schedule, what are the project funding requirements and are these in conflict?

Organisation – how is the client organisation structured and what are the key activities and processes that would impact the project? From which departments will the client representatives be drawn, how is the project to be run by the client decision unit, how will this interface with the project design and construction teams and what communication networks will be used for controlling the project? Ideas also about the core and non-core business of the organisation should be encompassed and how they relate to space use if a building project is under study at a workshop.

Schedule/time – what is the project duration, what is the anticipated duration for each stage of the project, what are the time constraints and from where do these derive? What is the relationship – if any – between the project schedule and funding arrangements?

People, skills and expertise – will personalities have an influence on the development of the project? What disciplines or expertise will be required and does this pose any problems?

Contractual issues – how are the parties to be linked contractually, with which forms of contracts and at what time are they to be used?

Project concept – can the project concept be easily fixed at a point in time or is it likely to evolve and change? If so why, and can this be altered? What influence will this have on the project as it develops over time?

Stakeholder analysis – customers, user groups or stakeholders may have different and perhaps conflicting interest in the project. Who are the key stakeholders and will their views impact project development in a different way?

Project constraints – what are they, are they real or apparent and what are their origins?

16.1.2 Interviews

It is advisable for the facilitator(s) to interview the participants of the VM workshop and the major stakeholders of the project in question (who should really participate). To aid the interviewing the checklist detailed in section 16.1.1 may be used. The aim of the interviews is to give the facilitator an overview of the strategic and tactical issues surrounding the project and to allow the first identification of mismatches. The interviews will lead to an appropriate agenda and the initial selection by the facilitator of the tools and techniques to be used in the workshop.

16.1.3 Document Analysis

In a similar way and for similar reasons to interviewing, the facilitator will gather all available documents as provided by the client and the various participants. These may include the client's business case detailing issues such as capital and revenue, operating expenses, taxation, financing and cash flow. Documents will also include previous reports, including VM workshop reports. The project manager's updated project execution plan will also be a useful document. It is advantageous for the facilitator to 'read into' the project by reviewing the correspondence files to date.

16.1.4 Questionnaire

In the event that interviews are not practical, a questionnaire may be sent to the participants before the workshop. The objectives will be same as for interviewing.

16.1.5 Site Tour

A visit to the site by the facilitator(s) and also, if possible, the participants will be advantageous in understanding the physical context of the project. In the absence of a site tour, photographs may achieve the same function.

16.1.6 POE/PPE Reports

Post-occupancy evaluation or post-project evaluation reports of similar projects provide a valuable source of information and should be obtained or carried out wherever practicable.

16.1.7 Facilities Walk-Through

A walk through of a similar facility of the client can assist in understanding space and arrangement.

16.1.8 Assembly of Production Drawings and Contract Programme

Following the appointment of the contractor, by whatever procurement method, control must be maintained through an accurate drawing register. It is necessary to obtain for the workshop all drawings pertaining to the contract, produced by both the design team and the production team. A full contract programme is also required.

16.1.9 Identification of Major Work Packages

This exercise is undertaken during interviews with the contractor's production team and results in a preliminary statement of the grouping of elements of the works. It is a necessary precursor to the supply chain exercise described in section 16.2.18.

16.1.10 Change Management Procedures

Any change to the works must be acceptable to the client, since this is the only party that can sanction change. Any change to the contract must be acceptable to both client and contractor. During interviews with the client and the client's consultants it is necessary to make explicit the procedures necessary for the workshop to carry through change proposals. Any value management change proposals (VMCPs) should aim to have the characteristics of:
- reducing the project programme
- not delaying the project programme (subject to client's value criteria)
- abortive costs of redundant design and the costs for the new design are paid for out of the savings generated by the VMCP (self-funding)
- generating savings for both client and contractor
- reducing risk for client and contractor.

16.1.11 Understanding the Contractor's Viewpoint and Strategy

The contractor, prior to pre-qualification, will decide how important the prospective contract is in terms of turnover and future workload. This analysis of workload will determine the contractor's tender strategy.

The contractor will have reviewed where the risk issues are during the tender process, and the possible exposure in undertaking the project works. The contractor will also look at opportunities to:
- gain competitive advantage
- reduce risk during the project works
- reduce reliance on specialist contractors or other resources
- provide ease of construction
- reduce costs
- deliver the project early (if it is perceived as a benefit)
- minimise temporary works (ensure permanent works design acts in lieu of temporary works).

In the process of establishing the contractor's value criteria there will be some linkage to the client's value criteria. However, the contractor can only assess the latter through the analysis of the project/contract documentation in terms of time, cost and quality and this may result in a misunderstanding of client preferences. Although the contractor may perceive the primary value criteria, there may be others present that are not apparent or of equal importance.

A major aspect of the value management study is the alignment of the contractor's and client's value systems.

16.2 Workshop Information Phase Tools

16.2.1 Presentation and Team Building

Short introductions are made by the individual workshop participants, indicating their part in the project and the information they have brought, or

been asked by the facilitator to bring, to the workshop. If necessary, based on whether the client is new to the concept of VM, the facilitator may give a short introduction to the principles of VM and an explanation of the agenda. It is also advisable for the sponsor (client) of the VM process to give a short presentation setting out the goals and aims of both the project and the VM process.

The object of these presentations is to pose the questions:
- Why is the workshop taking place?
- What is to be accomplished in the workshop?
- What does the client or end user need and/or want?
- What is the client's value system (usually expressed in terms of time, cost and quality)?
- What is the strategic and policy thinking behind the project or facility?

16.2.2 Brainstorm Issues, Group and Theme
A very simple technique is for the workshop participants to brainstorm relevant issues and write them down on sticky notes. The team then groups the issues under headings on the wall. Different coloured dots can be used to highlight important issues: for example, red dots may be used to indicate project show-stoppers. The technique very quickly highlights critical areas and helps the team to focus. It is also a good knowledge leveller and puts team members' previous assumptions into context. It takes between half an hour and an hour to complete the themes, groups and priorities. Critical areas are then reviewed and explored in more depth.

16.2.3 Stakeholder Analysis
Customers, user groups or stakeholders may have a different and perhaps conflicting interest in the project. The workshop participants should review their impacts on the project.

16.2.4 Strategic Time Line
This is a graphical, interactive, forcing and closing technique which establishes what needs to occur when. It is performed after the brainstorming issues, group and theme procedure and takes about half an hour to complete.

16.2.5 Project Driver Analysis
Project driver analysis identifies factors and/or people that are actively promoting the project. It is a forcing and closing technique, which establish six to ten key 'pushers' or 'pullers' of the project and precedes the time/cost/quality triangle. It generally takes between 10 and 15 minutes to perform. Problems may be encountered if a powerful project driver is present that is inhibiting project delivery.

16.2.6 Facilities Walk-Through
This is a mental facilities walk-through. The objective is for the workshop team to role-play key users of the facility and to get them to think through the ways in which people actually or could use the facility.

16.2.7 Time/Cost/Quality Analysis
The time/cost/quality comparison using a triangle approach is frequently used to establish the priorities of the project as a whole. By forcing the client to consider these three issues in a structured manner project priorities can be established. For example, figure 16.1 shows a T/C/Q triangle in which the client has placed emphasis on cost and quality.

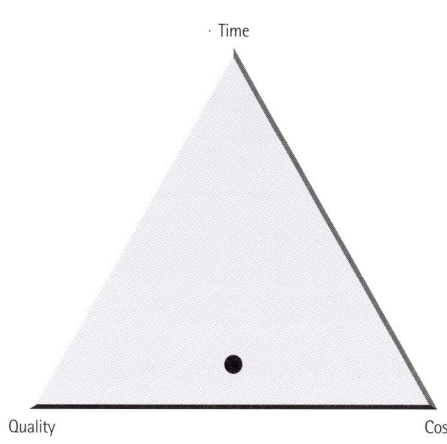

Figure 16.1 Example of a T/C/Q triangle

It is a good information-sharing and -synthesising technique that leads into a review of procurement strategies and options. It generally takes from five minutes to half an hour to conduct the analysis. The technique has several advantages: for example it is simple, it is very visual, it also exposes differences in perception between the participants and it stimulates information exchange. It is not uncommon for multiple triangles to emerge as part of this process, indicating conflicting priorities.

16.2.8 Function Analysis

Function analysis can be performed at various levels of the project life cycle. The procedure for establishing the necessary functions is outlined below.

Function analysis rules are:
1.) Verb/noun definition – most texts recommend that the function of an item or a system be expressed as a concise a phrase, ideally one comprising just an active verb followed by a noun.
2.) Functional definition/technical solution – a component or element represents a technical solution to a problem.
3.) Primary/secondary functions – primary functions are defined as those without which the project would fail or the task would not be accomplished. Secondary functions, on the other hand, are those that are a characteristic of the technical solution chosen for the primary function and are not required.
4.) Cost/worth – cost is the price paid or to be paid. Worth is the optimum cost to perform the required function or the cost of the optimum cost functional equivalent.

It is useful to ask the questions:
- what must the project, process, activity, elements and components do?
- what should they do?

From the above questions, it is then possible to:
- compile a list of verb/noun functions
- distinguish between the needs and wants functions (see section 5.2.1)
- finalise a list for construction of a function logic diagram, or move on to the creativity phase.

The relationship between value, function and cost is illustrated overleaf in figure 16.2.

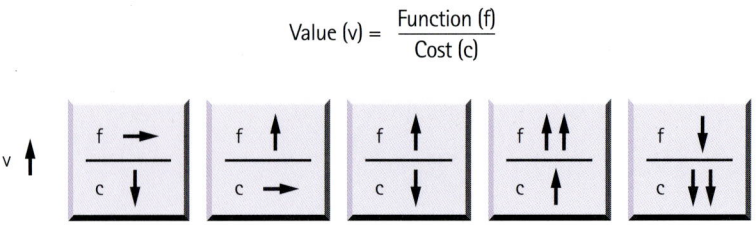

$$\text{Value (v)} = \frac{\text{Function (f)}}{\text{Cost (c)}}$$

Figure 16.2 The five connotations of function with cost giving an increase in value

16.2.9 The Function Logic Diagram

As an aid to information gathering and synthesis it is recommended that the workshop team should construct a function logic diagram. This will act as a knowledge leveller and is needed to establish consensus before progressing further.

The function logic diagram is used to determine the project task and why it exists, it is a very good deepening and focusing technique that describes a project extremely well. It brings together the value criteria of a project. The strategic issues that are highlighted represent the client's value system.

It takes between one and three hours to complete. Functions are brainstormed, then grouped and structured into a hierarchy. There is an expansion taking place during brainstorming which is then channelled and focused with the structuring of the hierarchy. Adding weights, costs and possibly time to the resulting diagram may be advantageous to highlight critical functions. The

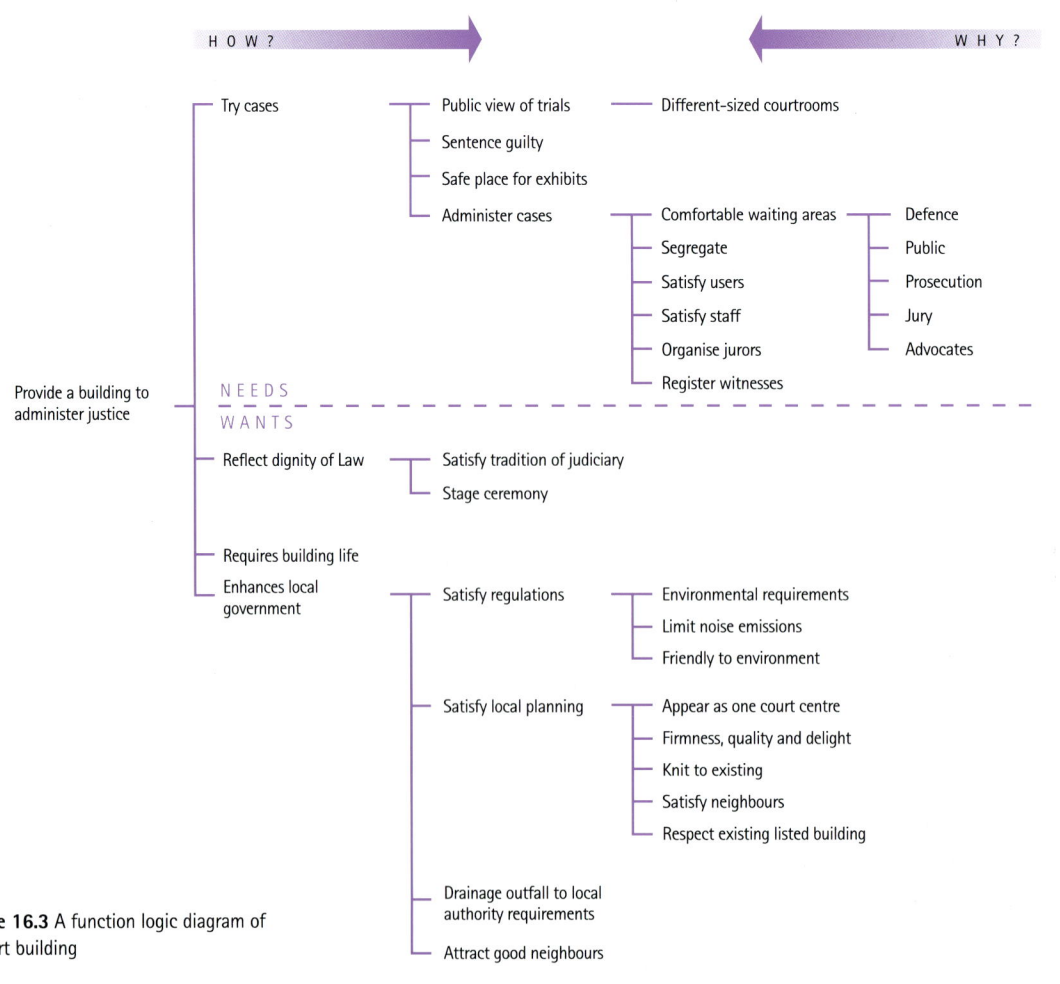

Figure 16.3 A function logic diagram of a court building

benchmarking study indicated that costing or weighting is the normal procedure for those who use function logic diagramming, but this may be time-consuming. The important issue is to generate functions, group them and structure them into a hierarchy.

The function logic diagram technique needs to be carefully explained to the participants, both before and after the exercise is carried out. Not everyone is comfortable with the technique initially. Sometimes participants cannot see its use immediately after it has been created and this may hamper the rest of the process. However, it acts as a constant source of subconscious reference during the remainder of the workshop and provides an anchor for the full workshop process.

A function logic diagram is described as having a primary function representing why the project exists. Supporting functions represent client needs and wants. In a function logic diagramming exercise a concept is taken and developed from left to right (figure 16.3).

Once the hierarchy has been established it is useful to pose the question 'what if we go one level higher?' and then establish the resulting function or objective. This will force the participants to step 'back' from the project and place it in a wider contex, i.e. linking it into corporate and business strategies.

16.2.10 Spatial Adjacency Analysis

A spatial adjacency analysis is a 'mileage chart' diagram such that each space can be related to another. The inter-relationship of the spaces is rated by the workshop team, usually using a score of between −5 and 5 according to how close or far apart the various spaces should be to one another in an optimal design. A score of 5 indicates close proximity is required, whereas a score of −5 indicates spatial adjacency is to be avoided. For example, in the case of a courtroom it may be important that the courtroom and judge's robing room are close together. This spatial relationship could score a 4.

It is a good comparative tool and highlights differences between the ideal and actual designs; solutions can then be suggested to solve mismatches. The technique is very dynamic and is excellent for prioritising and structuring proximity during briefing and for testing the existing design in a review project.

16.2.11 REDReSS

A REDReSS exercise may be carried out. This aims to align any previous work correctly with the client's view of certain standard criteria:

R   Re-organisation – addresses the extent to which the client's organisation is likely to undergo technological or organisational change during the design period or immediately thereafter.

E   Expansion – the likelihood of expansion of the organisation in the short and medium term that may impact immediately on the design, and in the long term on space external to the project.

D   Disposal – at the end of the project's physical or economic life will the client sell or demolish and re-build?

Re   Refurbishment and maintenance – does the client have a policy or a requirement for refurbishment?

S   Safety – projects can meet the safety requirements of regulations in many ways.

S   Security – to be taken into account in most projects.

This, like all value management exercises, helps the client and the team to focus on relevant issues and their associated problems. It is a good summarising and closing technique that raises whole life cycle issues, highlights safety and security, and also focuses on reorganisation and maintenance. Generally it takes 10 to 30 minutes to conduct. However, the technique assumes forward planning which may not have been overtly expressed previously.

16.2.12 SWOT Analysis

This is a technique that can be used at various steps during the workshop process to analyse the strengths, weaknesses, opportunities and threats (SWOT) of the design.

16.2.13 Process Flow Charting

This technique is generally used in briefing workshops as shown in figure 3.1. Most projects or facilities involve a process, are constructed to accommodate a process or are designed to facilitate that process. The idea here is to identify the key components making up processes and representing them as a flow chart diagram. This will ensure that the workshop team fully understands the technology involved in the organisation.

User flow diagramming, which is a variation of process flow charting, is a simple technique where flow charts are used to establish how each user acts outside and within a building. It is a technique for briefing space. It helps the team to understand space, linkages between spaces and users of space. The technique may be used as either a briefing tool or as a design review tool.

User flow diagramming is a fast way of briefing or reviewing space and generally takes between one and one-and-a-half hours to complete. It highlights space activities and communications requirements, and relates users and space. Core and non-core users are also identified. Care needs to be taken, however, so that it does not become too complex and the team start to lose sight of the important issues.

16.2.14 Value vs. Cost

Generally used on OSD and FSD workshops as shown in figure 3.1. The value/cost technique aims to rank the functions that have been identified according to their perceived value and their estimated cost. Four possibilities exist:
1. high value/high cost functions
2. high value/low cost functions
3. low value/high cost functions
4. low value/low cost functions.

The functions categorised under items 2 and 3 would be targeted for a further workshop.

16.2.15 Histogram of Cost

Generally used on OSD and FSD workshops where accurate costs are available, as shown in figure 3.1. Areas of the project are costed, a histogram is drawn up and the Pareto rule is then used to identify where to focus attention during the subsequent creativity phase. It is also a useful tool to compare the project cost under study with similar projects that have already been completed, in order to identify any mismatches.

This technique is also used during an operations workshop but is then based upon more accurate tender information rather than a pre-contract estimate or cost plan.

16.2.16 Element Function Analysis

The Building Cost Information Service defines an element as being that part of construction which fulfils the same function irrespective of the components from which it is made, e.g. a window or an external wall. An element function analysis commences by generating a list of functions being characteristically performed by a particular element. The element function list is then particularised to the design by eliminating those functions which are not relevant to the project under review.

16.2.17 Component Analysis

Key components are analysed in terms of their functions, answering the questions:
- what does it do?
- what does it cost?
- what else would perform the function?
- what does that cost?

The results are always selected without prejudice to quality and function. Component analysis is more likely at the FSD and operational stages.

16.2.18 Supply Chain Analysis

This involves identifying the key suppliers, when they are required to be involved on the project, their cost and whether there are any alternatives available to meet the required functions. Supply chain analysis can involve both component as well as labour inputs or could even encompass specialist design. The analysis may also consider opportunities to involve members of the supply chain in making contributions to the design, operational sequencing and delivery profiling as well as other potential technical alternatives.

16.2.19 Functional Specification

A functional specification is the functional analysis of a product from the customer's perspective. The objective is to determine the functions and quality required by the customer and the provision of these in the most effective manner. For example, customer A may only require a cladding system which keeps out the weather, whereas customer B may require one which keeps out the weather *and* makes an aesthetic statement. There are obviously different design, construction and cost implications due to these different functional requirements. The data for this description have been gathered from manufacturers and in a construction context may require these inputs much earlier in the process, where a key supplier is involved.

16.2.20 Pinch Point and Committal Point Analysis

At particular points in the project delivery process decision points arise at which outcomes are irreversible, commit substantial resources or have major time and cost implications to reverse. One example is the submission of a planning application. It is important that these points are recognised as a part of the value opportunity process.

16.3 VM Workshop Creativity Phase Tools

16.3.1 Brainstorming

Brainstorming is the most popular and well-known technique and has been found to be the most effective way of quickly harnessing team creativity. The basic principles of brainstorming are:

- the quantity of ideas is important regardless of how wild they are
- no criticism of ideas is allowed
- team participants should be encouraged to build on or modify ideas
- suggestions outside disciplines or expertise are encouraged
- no evaluation of ideas should be permitted; evaluation is conducted later during the judgement phase.

Brainstorming is almost exclusively the preferred technique during the creativity phase, mainly due to its simplicity, effectiveness and short duration (about an hour). Ideas are generally characterised by:

- strategic and business issues
- project management and time/cost/quality issues
- design and technical issues
- risk issues.

It should be noted that brainstorming of solutions should ideally never be conducted on the same day as the functional analysis as people are psychologically exhausted after the functional analysis exercise; there is also a real need to walk away and let thoughts settle.

Other creativity techniques do exists; further information may be obtained in Kelly and Male (1993), Norton and McElligott (1995) and Parker (1985).

16.3.2 Brainstorming Risk

Generating ideas may establish areas of risk and these should be identified and explored by:

- categorising into low, medium and high
- establishing potential impact as low, medium and high
- exploring interaction effects.

Explore management options for medium and high risks and impacts.

16.4 VM Workshop Evaluation Phase Tools

16.4.1 Technically Feasible / Economically Viable / Functionally Acceptable / Acceptable to Client

This is a non-numerical technique for evaluating ideas by discussing whether or not they conform to each of the above criteria. It is a very quick technique that is readily supported by the workshop team. Normally, ideas that have four ticks under the above headings are worth developing in outline. Those with less than four ticks may either require further research or be discarded on a second pass.

16.4.2 Silence Means 'No'

This is a very quick technique for rapidly discarding less feasible ideas. The facilitator runs through all the ideas with the team and deletes those that no one speaks out for, i.e. for which there is no case for keeping.

16.4.3 'Coloured Dots'

A non-numerical and very quick technique whereby team members are given a fixed number of coloured labels, which they stick to ideas they believe worthy of development. Those ideas attracting the greatest number of labels will probably have the greatest potential for development.

16.4.4 Championing

People volunteer or are chosen because of their suitability to be the champion of an idea or several ideas. This, in effect, means that they are responsible for seeing the idea through to its implementation or alternatively for justifying why it cannot be implemented. Championing has been found to increase the rate of implementation as people are made accountable for their actions with regard to the idea(s).

16.4.5 Big Issues

The ideas having the greatest impacts are chosen for implementation. The Pareto rule (80/20) may be used as a guide when selecting which ideas to bring forward for implementation. This may be used in conjunction with championing and really depends on the choice of the client and how much time is available.

16.4.6 Decision Matrices

These are numerical means of judging ideas by assigning weight values according to their impact and importance on the project. They are generally used when several solutions exist for one function. This is a useful and robust technique which is also useful for:
- choice of procurement strategy
- choice of contract strategy
- choice of appropriate site.

Principles for using decision matrices are:
- producing criteria to evaluate options
- weight criteria
- scoring options against criteria
- comparing and contrasting weights and scores
- exploring highest scored options against cost to determine cost/worth.

Generally, the workshop team is willing to accept the outcome and reaches consensus. However, it is time-consuming to carry out the exercise. The technique is described in ICE (1996), CIRIA (1996) and Norton and McElligott (1995).

16.5 VM Workshop Development Phase Tools

16.5.1 Establishment of Project Mission and Outline Specification

The development phase of a pre-brief workshop involves taking a number of the ideas for project development and auditing these against the mission for the project as realised through the function diagram. This confirms the project mission and takes forward the ideas for development into an outline or framework specification for the project.

16.5.2 Detailed Functional Specification

The development phase of a brief or Charette workshop involves taking a number of the ideas for project development and auditing these against the mission for the project as realised through the function diagram. This confirms the project mission and takes forward the ideas for development into a detailed functional specification for the project which will become the project brief. In the context of the Charette the existing brief may need to be revised to address points raised during the workshop.

16.5.3 Outline Time/Cost/Quality Analysis

During the development phase the ideas generated are developed as outline drawings that are produced to describe the project in sufficient detail to apply, for example, for planning permission. The outline drawings will allow the production of the first cost plan and time line programme. This part of the development stage is commonly undertaken in the workshop only to the stage which allows the design team to understand the proposed idea in sufficient detail to develop it in their own office.

16.5.4 Detailed Time/Cost/Quality Analysis

During the development phase of a final sketch design workshop the ideas generated are generally at an elements and components level. These are developed as detailed drawings that allow the production of the revised cost plan and time line programme. This part of the development stage is commonly undertaken in the workshop only to the stage which allows the design team to understand the proposed idea in sufficient detail to develop it in their own office.

16.5.5 Life Cycle Costing

Life cycle costing (LCC) may be defined as a technique for economic evaluation that accounts for all relevant costs during the investor's time horizon and adjusting for the time value of money. Carrying out a life cycle costing exercise can be very time-consuming and is therefore not always carried out within an actual workshop.

The costs which should be taken into account in a life cycle cost calculation are:

1. Investment costs – these include:
   - site cost
   - design fees (architect, QS, engineer, etc.)
   - legal fees
   - building costs
   - tax allowances: capital equipment allowance, capital gains, corporation tax, etc.
   - development grants.
2. Energy costs – heating, lighting, air conditioning, lifts, etc.
3. Non-energy operation and maintenance costs – these include:
   - letting fees
   - maintenance (cleaning and servicing)
   - repair (unplanned replacement of components)
   - caretaker
   - security and doormen
   - insurances
   - rates.
4. Replacement of components (planned replacement at end of useful life).
5. Residual or terminal credits – in determining these credits in the context of a building it is necessary to separate the value of the building from the value of the land. Generally, land appreciates in value but buildings depreciate until they become either economically or structurally redundant.

Some costs are not relevant and are not accounted for in the calculation. These costs are either trivial in amount or do not affect the decision.

16.5.6 Preparing and Making Presentations

The refined ideas supported by drawings, calculations and costs, where appropriate, must be presented by the workshop team to the body which commissioned the value management exercise. If a continuous workshop (where the various steps of the VM workshop process have been carried out in an uninterrupted sequence) has been held, then a presentation is a very good way to end. It helps the participants to refocus on the main issues, ideas and solutions of the workshop. These presentations form the basis for the proposals to be implemented as a consequence of the workshop.

16.6 VM Workshop Action Planning Phase Tools

16.6.1 Plan For Implementation

How to implement workshop proposals needs careful consideration and the following planning is necessary:

- define the responsibilities of the workshop team members with regards the proposals
- define the deliverables of the proposals
- establish priorities
- establish a realistic timetable for implementation of workshop proposals
- establish a review process to check on the success of proposal implementation.

16.6.2 Preparing Action Plan

The above checklist for 'plan for implementation' is used as the basis for preparing an action plan to be incorporated in the final workshop report. This action plan acts as an audit to check on post-workshop progression of proposal implementation.

16.6.3 Sign-off by Workshop Participants and Senior Management

To finalise a continuous workshop all the participants and senior management sign a document detailing that they agree with the workshop findings. In the case of workshops where development has taken place outside the workshop the final report will be available and should be signed by all parties. Advantages of a sign-off include:

- greater implementation
- greater team focus
- no scope for changing one's mind later
- fewer disagreements further into the project life cycle
- accountability by all parties.

16.7 Workshop Report

16.7.1 Preparation of Report

Report preparation is generally the responsibility of the value management facilitator. The report will normally describe the main items for improvement and an action plan with tasks assigned to particular workshop team members. The production of the report is enhanced if a full-time recorder is present at the workshop.

16.7.2 Circulation of report

The final report should be circulated to team members as soon as possible after the workshop. In this respect it is not uncommon for workshop members to take away the action plan or at least be sent this element of the report by fax on the following day.

16.8 Workshop Outputs

16.8.1 Statement of Project Mission

The statement of project mission is the reason for undertaking the project, where project is defined as 'the investment of resource for return'. The project mission is the performance statement, usually found at the extreme left of the function diagram, that incorporates the key elements of the client's value system. The project mission is derived before the decision to construct is taken and therefore the mission should not be expressed in terms of a construction solution, e.g. a new car showroom or a water treatment plant.

16.8.2 Performance Specification (Brief)
The output of the briefing study is the project brief. This style of workshop offers an alternative method of project briefing and results in a clear statement of not only the project specification but also the client's value system.

16.8.3 Outline Sketch Design (OSD)
The outline sketch design reflects the earliest design solution that expresses in drawn form a solution or a number of solutions to the project brief. OSD drawings are often produced by designers during the development phase of the workshop as a means of communicating ideas to the client at the presentation.

16.8.4 Final Sketch Design (FSD)
The final sketch design represents the last opportunity for changes to be made to the design and/or the client's requirements without incurring major costs. Once the final sketch design has been agreed the design team will commence the detailed design or the same task will be undertaken by a design/build contractor.

16.8.5 Decision to Build
The decision to build is taken by the client in the full knowledge of the mission statement and with reference to all of the facts explored during the information and development phases of the workshop.

16.8.6 Project Execution Plan (PEP)
The project execution plan is drawn up immediately following the pre-brief workshop. The PEP is a dynamic document updated at key points during the project life cycle. It should contain:
- the project mission
- the aims and objectives of the PEP
- the procedures for updating the PEP
- the project organisation structure of the client
- the consultants and their responsibilities
- the contractors, management contractor, construction management organisation
- descriptions of the project in performance terms incorporating the client's value system
- predicted cost and cash flow
- project reporting procedures and particularly the procedures for information distribution between the client's project team, consultants and contractor
- a schedule of key meetings and workshops (including value and risk workshops).

16.8.7 Production Programme
A time line is often produced at a workshop to focus the workshop team on major events. The time line is used to validate the production programme.

16.8.8 Workshop Report
The workshop report will generally comprise:
- an executive summary
- an action plan
- the primary results from the workshop
- final comments and conclusions
- an appendix containing a list of the workshop participants and all of the outcomes of the various workshop stages.

17. References

Building Research Establishment. *Value From Construction. Getting Started in Value Management.* BRE, 1997.

Construction Industry Board, *Briefing the Team.* Thomas Telford, 1997.

Construction Industry Research and Information Association. *A Client's Guide to Value Management in Construction*, prepared by Davis Langdon Consultancy and University of Reading, CIRIA, 1996.

European Commission. *Value Management Handbook.* 1995.

Hayden, G.W. and Parsloe C.J. *Value Engineering of Building Services.* BSRIA, 1996, Application Guide 15/96.

H M Treasury. *Value Management Guidance.* Central Unit on Procurement, 1996.

Institution of Civil Engineers. *Creating Value in Engineering.* Design and Practice Guide. Thomas Telford, 1996.

Kelly, J. R. and S. P. Male. *Value Management in Design and Construction: The Economic Management of Projects.* London, E & FN Spon, 1993.

Latham, M. *Constructing the Team.* HMSO, 1994.

Morris, P. W. G. and Hough, G. H. *The Anatomy of Major Projects: the Reality of Project Management.* Wiley, Chichester, 1987.

Norton, B. R. and McElligott W. C. *Value Management in Construction: A Practical Guide.* MacMillan Press, London, 1995.

O'Donnell, S. An Introduction to Group Decision Making and Group Decision Support Systems. *BT Technology Journal,* 1994, **12** (4).

Parker, D. E. *Value Engineering Theory.* McGraw Hill, New York, 1985